Permanents

GIAN-CARLO ROTA, *Editor*

ENCYCLOPEDIA OF MATHEMATICS AND ITS APPLICATIONS

Volume 6

Volume 1 LUIS A. SANTALÓ
Integral Geometry and Geometric Probability, 1976

Volume 2 GEORGE E. ANDREWS
The Theory of Partitions, 1976

Volume 3 ROBERT J. McELIECE
The Theory of Information and Coding, 1977

Volume 4 WILLARD MILLER, Jr.
Symmetry and Separation of Variables, 1977

Volume 5 DAVID RUELLE
Thermodynamic Formalism, 1978

Volume 6 HENRYK MINC
Permanents, 1978

Other volumes in preparation

ENCYCLOPEDIA OF MATHEMATICS and Its Applications

GIAN-CARLO ROTA, Editor
Department of Mathematics
Massachusetts Institute of Technology
Cambridge, Massachusetts

GIAN-CARLO ROTA, *Editor*

ENCYCLOPEDIA OF MATHEMATICS AND ITS APPLICATIONS

Volume 6

Section: Linear Algebra
Marvin Marcus, *Section Editor*

Permanents

Henryk Minc

Department of Mathematics
University of California
Santa Barbara, California

With a Foreword by

Marvin Marcus

Department of Mathematics
University of California
Santa Barbara, California

1978

Addison-Wesley Publishing Company
Advanced Book Program
Reading, Massachusetts

London · Amsterdam · Don Mills, Ontario · Sydney · Tokyo

Library of Congress Cataloging in Publication Data

Minc, Henryk.
Permanents.

(Encyclopedia of mathematics and its applications;
v. 6: Section, Linear algebra)
Bibliography: p.
Includes index.
1. Permanents (Matrices) 2. Inequalities (Mathematics) I. Title. II. Series.
QA188.M56 512.8'43 78-6754
ISBN 0-201-13505-1

Published by the Press Syndicate of the University of Cambridge
The Pitt Building, Trumpington Sreet, Cambridge CB2 1RP
32 East 57th Street, New York, NY 10022, USA
Stamford Road, Oakleigh, Melbourne 3206, Australia

me (1970):

First published 1978 by Addison Wesley
First published by Cambridge University Press 1984

Printed in the United States of America

ISBN 0 521 30226 9

Contents

Editor's Statement

A large body of mathematics consists of facts that can be presented and described much like any other natural phenomenon. These facts, at times explicitly brought out as theorems, at other times concealed within a proof, make up most of the applications of mathematics, and are the most likely to survive changes of style and of interest.

This ENCYCLOPEDIA will attempt to present the factual body of all mathematics. Clarity of exposition, accessibility to the non-specialist, and a thorough bibliography are required of each author. Volumes will appear in no particular order, but will be organized into sections, each one comprising a recognizable branch of present-day mathematics. Numbers of volumes and sections will be reconsidered as times and needs change.

It is hoped that this enterprise will make mathematics more widely used where it is needed, and more accessible in fields in which it can be applied but where it has not yet penetrated because of insufficient information.

Ordinarily, a specialized volume, such as this one, would call for a prior work on determinants. It is hoped that, in time, a companion volume will be added to the ENCYCLOPEDIA. Meanwhile, Professor Minc's monograph is expected to remain the definitive treatment on permanents and, as the author wittily remarks, the only one in all probability.

A permanent is an improbable construction to which we might have given little chance of survival fifty years ago. Yet the numerous appearances it has made in physics and in probability betoken the mystifying usefulness of the concept, which has a way of recurring in the most disparate of circumstances. Thanks to Professor Minc's efforts, they are now all collected here.

Gian-Carlo Rota

Foreword

In referring to Sir Thomas Muir and his monumental work *The Theory of Determinants in the Historical Order of Development*, Minc calls him the "master from Edinburgh." As a graduate of that venerable institution himself, Minc carries on with this high tradition of scholarship and masterly exposition with *Permanents*. The permanent function has been studied for more than a century. As Minc amusingly points out in Section 1.1, the word "permanent" originated with Cauchy in 1812, although a referee of one of Minc's earlier papers admonished him for daring to invent such a ludicrous name.

In the Carus monograph *Combinatorial Mathematics*, H. J. Ryser mentions that the permanent "appears repeatedly in the literature of combinatorics in connection with certain enumeration and extremal problems." As an example, if D is the n-square matrix with 0's on the main diagonal, 1's elsewhere, then per(D) is a count of the total number of derangements—that is, permutations with no fixed points—of $1,\ldots,n$. The Laplace expansion theorem works equally well for permanents as for determinants—indeed it is simpler, since no sign changes arise. From this it follows immediately that

$$\operatorname{per}(A+B)=\sum_{r=0}^{n}\sum_{\alpha,\beta}\operatorname{per}A[\alpha|\beta]\operatorname{per}B(\alpha|\beta), \tag{1}$$

where the inner summation is over all products of an $r\times r$ subpermanent of A lying in rows $\alpha=(\alpha_1,\ldots,\alpha_r)$, columns $\beta=(\beta_1,\ldots,\beta_r)$, and the complementary subpermanent of B. If we take $A=J$, the matrix of all 1's, $B=-I_n$, then $D=A+B$, and the number of derangements is given by the remarkable formula

$$\begin{aligned}\operatorname{per}(J-I_n)&=\sum_{r=0}^{n}\sum_{\alpha}\operatorname{per}J[\alpha|\alpha](-1)^{n-r}\\&=\sum_{r=0}^{n}(-1)^{n-r}\frac{n!}{(n-r)!}.\end{aligned}$$

Let U_n be the nth menage numbers; that is, U_n is a count of the number of permutations σ of $1, 2, \ldots, n$ such that $\sigma(i)$ is neither i nor $i+1 \pmod{n}$ $i=1,\ldots,n$. Analogously to the derangement problem we have

$$U_n = \operatorname{per}(J - I_n - P),$$

where P has 1's in positions (1,2), (2,3),...,$(n-1,n)$, $(n,1)$, and 0's elsewhere. Thus, as Ryser suggests, the permanent *function* is the "correct" tool for dealing with a number of difficult enumeration problems for restricted permutations. Minc is certainly a master in making the computations required for such problems (see Section 3.4). In fact, I have personally watched while Minc punched some quite remarkable permanents of circulants out of one of the more primitive hand-held calculators of the early sixties.

Although a number of deep and interesting results about the permanent have been obtained by direct methods, there is a somewhat oblique approach to the function that has proved to be quite productive over the last two decades. Let V be an n-dimensional inner product space. Then the Z-graded contravariant tensor space over $V, T_0(V) = C \dot{+} V \dot{+} V \otimes V \dot{+} V \otimes V \otimes V \dot{+} \ldots$, inherits an inner product from V that satisfies the formula

$$(x_1 \otimes \ldots \otimes x_p, y_1 \otimes \ldots \otimes y_p) = \prod_{t=1}^{p} (x_t, y_t)$$

for homogeneous decomposable elements of degree p. The symmetric space, $\dot{V}$, is the range in $T_0\,(V)$ of the symmetry operator

$$\sum_{p=0}^{0} \mathbb{S}_p;$$

$\mathbb{S}_p = \dfrac{1}{p!} \Sigma\sigma$, and the summation is over the symmetric group of degree p (the action of σ on a decomposable tensor is defined by $\sigma(x_1 \otimes \ldots \otimes x_p) = x_{\sigma(1)} \otimes \ldots \otimes x_{\sigma(p)}$). Each $\mathbb{S}_p$ is a hermitian idempotent, so that, if $x_1 \ldots x_p = \mathbb{S}_p x_1 \otimes \ldots \otimes x_p$, we have

$$\begin{aligned} (x_1 \ldots x_p, y_1 \ldots y_p) &= (x_1 \otimes \ldots \otimes x_p, \mathbb{S}_p y_1 \otimes \ldots \otimes_p) \\ &= \frac{1}{p!} \sum_{\sigma} \prod_{t=1}^{p} (x_t, y_{\sigma(t)}) \\ &= \frac{1}{p!} \operatorname{per}((x_i, y_j)). \end{aligned}$$

Thus, the permanent function appears naturally as an analytical expression for the inner product in $V^{(p)} = \operatorname{im} \mathbb{S}_p$ in precisely the same way as the determinant does in the pth exterior space $\wedge^p V$. This means that the unitary geometry of $V^{(p)}$ is available for investigating per(A), and it is this observation that has led to substantial progress in dealing with the function. Minc skillfully interweaves the combinatorial and multilinear approaches to the function throughout the book.

Certainly *Permanents* is the definitive treatise. The history, theory, and applications are completely surveyed, and the bibliography contains a reference to every book and paper written on the subject. No doubt the present book will result in renewed interest in this intractable and fascinating matrix function.

MARVIN MARCUS
General Editor, Section in Linear Algebra

Preface

Permanents made their first appearance in 1812 in the famous memoirs of Binet and Cauchy. Since then 155 other mathematicians contributed 301 publications to the subject, more than three-quarters of which appeared in the last 19 years. The present monograph is an outcome of this remarkable re-awakening of interest in the permanent function.

The purpose of the book is to give a complete account of the theory of permanents, their history and applications, in a form accessible not only to mathematicians but also to workers in various applied fields, and to students of pure and applied mathematics. Here is the first complete account of the theory of permanents. It is a survey in the style of MacDuffy *The Theory of Matrices* and of *A Survey of Matrix Theory and Matrix Inequalities*, by Marcus and Minc. However, it differs from both works in several respects: the style is more leisurely, the proportion of theorems proved in the text is higher, and the scope is wider—the volume covers virtually the whole of the subject, a feature that no survey of the theory of matrices can even attempt. Apart from many theorems proved in detail, there are numerous results stated without proof. Due to limitation of space, not every known result could be mentioned in the text. The choice of the theorems included in the book reflects, of course, the author's predilections.

The first chapter of the monograph is a historical survey of the theory of permanents since its beginnings in 1812. Here only some classical results are discussed in detail. In Chapter 2 general properties of permanents are developed. Chapter 3 is devoted to combinatorial and structural properties of (0.1)-matrices. The next three chapters may be regarded as the heart of the monograph. They deal with inequalities involving permanents, and with lower and upper bounds for permanents. The latter are particularly important due to the lack of efficient methods for computing permanents. One of the three chapters, Chapter 5, contains an up-to-date survey of the literature on the famed van der Waerden conjecture. In Chapter 7 we discuss several methods for computing permanents and compare their efficiency. The concluding chapter contains a section on some important topics that do not fall under the headings of the preceding chapters, a section on applications of permanents to combinatorics, graph theory, and to statistical mechanics, and two sections in which we report on the present status of the conjectures and problems in the Marcus–Minc 1965 list, and compile a new list of unresolved conjectures and unsolved problems on permanents.

Every chapter concludes with a set of problems of varying difficulty. Thus the book can be used as a text for a course at the advanced undergraduate or graduate level. The only prerequisites are a standard undergraduate course in the theory of matrices and a measure of mathematical maturity.

A special feature of the monograph, and, in fact, its foundation is the Bibliography which contains every paper and book on permanents published before the end of 1977 or awaiting publication at that time. The Bibliography also includes some papers on cognate topics even if they make no explicit use of permanents but can be interpreted in terms of permanents. Thus several classical papers on the "problème des ménages" are listed. Papers on Schur functions are included if the specialization of the results to permanents produces new significant theorems. Articles on graphs and on combinatorial properties of matrices are excluded unless they are related to or make use of permanents. In general, only the most important result of papers are reviewed in the Bibliography. In case of papers or books covering more than one area, only the part related to permanents is reviewed.

The usual double numeration is used in references. Thus "Section 3.2" refers to Section 2 in Chapter 3. The fourth theorem in Section 3.2 is referred to as "Theorem 2.4, Chapter 3"; similarly with references to examples and exhibited formulas. Within a chapter the reference to the chapter is omitted; e.g., within Chapter 3 the above theorem is quoted simply as "Theorem 2.4".

I should like to express my appreciation to Mrs. Barbara Federman for her assistance in preparing the book, to the Director and the staff of the Institute for the Interdisciplinary Applications of Algebra and Combinatorics, U.C.S.B., for having the manuscript typed and assembled, and, in particular, to Mrs. Michelle Dunn for her excellent job of typing. The work on the book was supported in part by the Air Force Office of Scientific Research under Grants AFOSR-72-2164 and AFOSR-77-3166.

HENRYK MINC

CHAPTER 1

The Theory of Permanents in the Historical Order of Development

1.1 Introduction

Modern mathematicians have a proclivity to invent flippant names for newly introduced mathematical entities and concepts. They delight in talking about mobs, radicals, derogatory matrices, osculating planes, improper ideals, etc. It may appear that the term "permanent" was also invented by a waggish algebraist. In fact, a few years ago a well-meaning referee admonished the author for daring to invent this ludicrous name for a function that Schur himself introduced without designating it by any specific term. The fact of the matter is that the permanent function was studied and called by that name before Schur was even born.

In his famous memoir of 1812, Cauchy [2] developed the theory of determinants as a special type of alternating symmetric functions, which he distinguished from the ordinary symmetric functions by calling the latter "fonctions symétriques permanentes." He also introduced a certain subclass of symmetric functions, which were later named *permanents* by Muir [14] and which are nowadays known by this name. These functions can be defined by means of matrices and modern notation as follows.

Let $A=(a_{ij})$ be an $m\times n$ matrix over any commutative ring, $m\leqslant n$. The *permanent* of A, written Per(A), or simply Per A, is defined by

$$\operatorname{Per}(A)=\sum_{\sigma} a_{1\sigma(1)}a_{2\sigma(2)}\cdots a_{m\sigma(m)}, \tag{1.1}$$

where the summation extends over all one-to-one functions from $\{1,\ldots,m\}$ to $\{1,\ldots,n\}$. The sequence $(a_{1\sigma(1)},\ldots,a_{m\sigma(m)})$ is called a *diagonal* of A, and the product $a_{1\sigma(1)}\cdots a_{m\sigma(m)}$ is a *diagonal product* of A. Thus the permanent of A is the sum of all diagonal products of A.

ENCYCLOPEDIA OF MATHEMATICS and Its Applications, Gian-Carlo Rota (ed.). Vol. 6: Henryk Minc, Permanents

ISBN 0-201-13505-1

For example, if

$$A=[3\,2\,4],$$

$$B=\begin{bmatrix}3 & 2 & 4\\ 2 & 1 & 5\end{bmatrix},$$

$$C=\begin{bmatrix}3 & 2 & 4\\ 2 & 1 & 5\\ -1 & 2 & -2\end{bmatrix},$$

then Per $A=9$, Per $B=44$, and Per $C=18$.

The special case $m=n$ is of particular importance. We denote the permanent of a square matrix A by per(A) instead of Per(A). In fact, most writers restrict the designation "permanent" to the case of square matrices.

1.2 The Originators: Binet and Cauchy

Permanents were introduced in 1812 almost simultaneously by Binet [1] and Cauchy [2]. Binet in his memoir also gave formulas for computing the permanents of $m\times n$ matrices for $m\leqslant 4$.

The permanent of an $m\times n$ matrix A, $m\leqslant n$, is the sum of all the diagonal products of A. In other words, Per A is the sum of all products of m elements of A, no two in the same row or the same column. It follows that all the terms of Per$A$, and many other superfluous terms, are contained in the set of terms obtained by multiplying the row sums of A. For example, if A is a $2\times n$ matrix, then

$$\operatorname{Per}A=\sum_{s\neq t} a_{1s}a_{2t},$$

while the product of the row sums of A is

$$\prod_{i=1}^{2}\sum_{j=1}^{n} a_{ij}=\sum_{s,t=1}^{n} a_{1s}a_{2t}$$

$$=\sum_{s\neq t} a_{1s}a_{2t}+\sum_{s=1}^{n} a_{1s}a_{2s}.$$

Hence

$$\operatorname{Per}A=\prod_{i=1}^{2}\sum_{j=1}^{n} a_{ij}-\sum_{s=1}^{n} a_{1s}a_{2s}, \tag{2.1}$$

which is Binet's formula for $m=2$.

ISBN 0-201-13505-1

Let $Q_{t,k}$ denote the set of all strictly increasing sequences of integers $\omega=(\omega_1,\ldots,\omega_t)$ satisfying $1\leqslant\omega_1<\omega_2<\cdots<\omega_t\leqslant k$. For an $m\times n$ matrix $A=(a_{ij})$ and a sequence $(i_1,\ldots,i_s)\in Q_{s,m}$, define

$$r_{i_1*\cdots*i_s}=\sum_{j=1}^{n} a_{i_1j}a_{i_2j}\cdots a_{i_sj}.$$

In particular,

$$r_i=\sum_{j=1}^{n} a_{ij}$$

denotes the ith row sum of A. Then Binet's formula (2.1) can be written in the form

$$\operatorname{Per}A=r_1r_2-r_{1*2}.$$

Now consider a $3\times n$ matrix $A=(a_{ij})$, $n\geqslant 3$. The product $r_1r_2r_3$ contains all the terms of Per$A$ and in addition $n^3-n(n-1)(n-2)$ "unwanted" terms such as $a_{11}a_{21}a_{3j}, a_{12}a_{22}a_{3j},\ldots,a_{11}a_{2j}a_{31},\ldots,a_{1j}a_{21}a_{31},\ldots;$etc., $j=1,\ldots,n$—that is, the terms of $r_{1*2}r_3$, $r_{1*3}r_2$, and $r_{2*3}r_1$. It seems, therefore, that if we subtract $r_{1*2}r_3+r_{1*3}r_2+r_{2*3}r_1$ from $r_1r_2r_3$, we should be left with the terms of Per$A$. Unfortunately, this is not the case. What happens is that, although we subtract all the "unwanted" terms, we subtract some of them more than once. To be precise, the terms $a_{1j}a_{2j}a_{3j}, j=1,\ldots,n$, appear in all three products $r_{1*2}r_3$, $r_{1*3}r_2$, and $r_{2*3}r_1$, and therefore each of them is subtracted three times instead of once. But r_{1*2*3} is the sum of all the $a_{1j}a_{2j}a_{3j}$. Thus we obtain Binet's second formula:

$$\operatorname{Per}A=r_1r_2r_3-(r_{1*2}r_3+r_{1*3}r_2+r_{2*3}r_1)+2r_{1*2*3}. \tag{2.2}$$

We now introduce the following simplifying notation: For $2\times n$ matrices,

$$S(1,1)=r_1r_2,$$
$$S(2)=r_{1*2};$$

for $3\times n$ matrices,

$$S(1,1,1)=r_1r_2r_3,$$
$$S(1,2)=r_1r_{2*3}+r_2r_{1*3}+r_3r_{1*2},$$
$$S(3)=r_{1*2*3};$$

ISBN 0-201-13505-1

for $4\times n$ matrices,

$$S(1,1,1,1)=r_1r_2r_3r_4,$$
$$S(1,1,2)=r_1r_2r_{3*4}+r_1r_3r_{2*4}+r_1r_4r_{2*3}+r_2r_3r_{1*4}$$
$$+r_2r_4r_{1*3}+r_3r_4r_{1*2},$$
$$S(2,2)=r_{1*2}r_{3*4}+r_{1*3}r_{2*4}+r_{1*4}r_{2*3},$$
$$S(1,3)=r_1r_{2*3*4}+r_2r_{1*3*4}+r_3r_{1*2*4}+r_4r_{1*2*3},$$
$$S(4)=r_{1*2*3*4},$$

etc. In general, if A is an $m\times n$ matrix and $t_1,\ldots,t_k$ are integers, $1\leqslant t_1\leqslant\cdots\leqslant t_k, t_1+\cdots+t_k=m$, then $S(t_1,\ldots,t_k)$ is the symmetrized sum of all distinct products of the $r_{i_1*\cdots*i_s}, s=t_1,\ldots,t_k$, so that in each product the sequences $(i_1,\ldots,i_s)\in Q_{s,m}, s=t_1,\ldots,t_k$, partition the set $\{1,\ldots,m\}$.

Using this notation we can write equations (2.1) and (2.2) in the following form:

$$\operatorname{Per}A=S(1,1)-S(2), \tag{2.1'}$$
$$\operatorname{Per}A=S(1,1,1)-S(1,2)+2S(3). \tag{2.2'}$$

Both these formulas were proved by Binet by a very involved method. He also gave, without proof, a formula for the permanent of a $4\times n$ matrix A:

$$\operatorname{Per}A=S(1,1,1,1)-S(1,1,2)+S(2,2)+2S(1,3)-6S(4). \tag{2.3}$$

This formula can be established by the use of the principle of inclusion and exclusion which we used to prove formulas (2.1) and (2.2).

Let A be an $4\times n$ matrix. The function $S(1,1,1,1)$ is the sum of all the terms of $\operatorname{Per}A$ and also some "superfluous" terms all of which are the terms of $S(1,1,2)$. We therefore subtract $S(1,1,2)$ from $S(1,1,1,1)$. However, we have "overreacted": Some of the terms, such as $a_{11}a_{21}a_{32}a_{42}$, $a_{11}a_{21}a_{31}a_{42}$, and $a_{11}a_{21}a_{31}a_{41}$, appear in $S(1,1,2)$ with multiplicity greater than 1. For example, $a_{11}a_{21}a_{32}a_{42}$ appears both in $r_{1*2}r_3r_4$ and in $r_1r_2r_{3*4}$. We compensate by adding $S(2,2)$. Now, we count the number of the terms of the form $a_{1i}a_{2j}a_{3j}a_{4j}, i\neq j$, in

$$S(1,1,1,1)-S(1,1,2)+S(2,2).$$

Each of them occurs once in $S(1,1,1,1)$, three times in $S(1,1,2)$ (for example, $a_{11}a_{22}a_{32}a_{42}$ appears once in each of $r_1r_2r_{3*4}$, $r_1r_3r_{2*4}$, and $r_1r_4r_{2*3}$) and does not appear in $S(2,2)$. Hence we compensate by adding twice $S(1,3)$:

$$S(1,1,1,1)-S(1,1,2)+S(2,2)+2S(1,3). \tag{2.4}$$

ISBN 0-201-13505-1

It remains to account for the terms $a_{1j}a_{2j}a_{3j}a_{4j}, j=1,\ldots,n$. Each of them appears once in $S(1,1,1,1)$, six times in $S(1,1,2)$, three times in $S(2,2)$, and eight times in $2S(1,3)$. Therefore we must subtract $1-6+3+8=6$ times $S(4)$ from (2.4), and the formula (2.3) follows.

Example 2.1. Compute the permanent of matrix

$$A=\begin{bmatrix} 1 & 1 & 0 & 1 & 1 \\ 0 & 1 & 1 & 1 & 1 \\ 1 & 0 & 1 & 0 & 1 \\ 1 & 1 & 0 & 1 & 0 \end{bmatrix}.$$

We compute:

$$S(1,1,1,1)=144, \qquad S(1,1,2)=151,$$
$$S(2,2)=13, \qquad S(1,3)=13, \qquad S(4)=0.$$

Hence,

$$\mathrm{Per}(A)=144-151+13+2\times 13-6\times 0=32.$$

Binet did not explain how he derived the coefficients in (2.4), nor did he give a general formula for the permanent of an $m\times n$ matrix for $m>4$.

Recently Binet's formula was generalized to any $m\times n$ matrices, $m\leqslant n$ [301]. This formula and a more efficient formula due to Ryser [87] will be given in the chapter on the evaluation of permanents.

1.3 The Continuators: Borchardt, Cayley, and the Master from Edinburgh—Sir Thomas Muir

During the century that followed the appearance of the memoirs of Binet and Cauchy, some twenty papers on permanents were published. Most of them dealt with identities involving determinants and permanents. The results that created the most interest are identities of Borchardt [4], Cayley [6], and Muir [14]. All three are formulas for the product of the permanent and the determinant of a matrix.

Let $A=(a_{ij})$ be an $n\times n$ matrix. Then

$$\mathrm{per}(A)\det(A)=\left(\sum_{\sigma\in E}\prod_{i=1}^{n}a_{i\sigma(i)}\right)^2-\left(\sum_{\sigma\in F}\prod_{i=1}^{n}a_{i\sigma(i)}\right)^2, \tag{3.1}$$

where E and F are the sets of even and odd permutations, respectively. The problem is how to express the difference on the right in a more

ISBN 0-201-13505-1

attractive form; for example, it is clearly equal to

$$\sum_{\sigma\in E}\prod_{i=1}^{n} a_{i\sigma(i)}^2-\sum_{\sigma\in F}\prod_{i=1}^{n} a_{i\sigma(i)}^2+f(A)=\det(A^{(2)})+f(A), \tag{3.2}$$

where $A^{(2)}=A*A$ is the matrix whose (i,j) entry is a_{ij}^2, and $f(A)$ represents the remaining terms. If $n=2$, then actually $f(A)=0$. For $n=3$, Cayley expressed $f(A)$ in terms of the determinant of a related matrix.

THEOREM 3.1 (Cayley [6]). *Let* $A=(a_{ij})$ *be a* 3×3 *matrix,* $a_{ij}\neq 0$, *and let* $A^{(-1)}$ *be the* 3×3 *matrix whose* (i,j) *entry is* a_{ij}^{-1}. *Then*

$$\operatorname{per}(A)\det(A)=\det(A^{(2)})+2\Big(\prod_{i,j}a_{ij}\Big)\det(A^{(-1)}). \tag{3.3}$$

The proof follows immediately from (3.1) and (3.2).

COROLLARY. *If* $B=(b_{ij})$ *is a singular matrix,* $b_{ij}\neq 0$, *then*

$$\operatorname{per}(B^{(-1)})\det(B^{(-1)})=\det(B^{(-2)}). \tag{3.4}$$

Here $B^{(-2)}$ denotes the 3×3 matrix whose (i,j) entry is b_{ij}^{-2}.

Borchardt obtained a formula similar to (3.4) for any n, but only for a special type of matrix.

THEOREM 3.2 (Borchardt [4]). *Let* A *be an* $n\times n$ *matrix whose* (i,j) *entry is* $(s_i-t_j)^{-1}$. *Then*

$$\operatorname{per}(A)\det(A)=\det(A^{(2)}). \tag{3.5}$$

We shall not offer a separate proof of Borchardt's result, since it is an immediate consequence of a generalization of Cayley's theorem by Carlitz and Levine [63], which we give at the end of this section.

Sir Thomas Muir occupies a unique position in the history of permanents, and even more so in the history of determinants. In his monumental *The Theory of Determinants in the Historical Order of Development* [25, 26, 35, 36, 40], he gives *inter alia* an abstract of every paper on permanents published before 1920, a third of which were his own contributions. Muir's papers deal mostly with expressions and identities involving permanents and determinants. Of these we give below one of the results in his first paper on permanents.

THEOREM 3.3 (Muir [14]). *Let* $A=(a_{ij})$ *and* $X=(x_{ij})$ *be n-square matrices. Then*

$$\operatorname{per}(A)\det(X)=\sum_{\sigma\in S_n}\varepsilon(\sigma)\det(A*X_\sigma), \tag{3.6}$$

ISBN 0-201-13505-1

where X_σ is the matrix whose ith *row is the* $\sigma(i)$th *row of* X, $A*X_\sigma$ *is the Hadamard product, and* $\varepsilon(\sigma)$ *denotes the sign of* σ.

(The Hadamard product of two $n\times n$ matrices, $P=(p_{ij})$ and $Q=(q_{ij})$, is the $n\times n$ matrix whose (i,j) entry is $p_{ij}q_{ij}$.)

Each side of (3.6) contains $(n!)^2$ terms: To each pair of permutations (φ,ψ) corresponds a term. Thus to (φ,ψ) we can make correspond on the left-hand side the term

$$\varepsilon(\psi)\prod_{i=1}^{n} a_{i,\varphi(i)}\prod_{t=1}^{n} x_{t,\psi(t)},$$

and to (φ,σ) we can make correspond on the right-hand side the term

$$\varepsilon(\sigma)\left(\varepsilon(\varphi)\prod_{i=1}^{n}(A*X_\sigma)_{i,\varphi(i)}\right)=\varepsilon(\sigma\varphi)\prod_{i=1}^{n} a_{i,\varphi(i)}x_{\sigma(i),\varphi(i)}$$

$$=\varepsilon(\varphi\sigma^{-1})\prod_{i=1}^{n} a_{i,\varphi(i)}\prod_{t=1}^{n} x_{t,\varphi\sigma^{-1}(t)}.$$

Now, $\psi\leftrightarrow\varphi\sigma^{-1}$ establishes one-to-one correspondence between the equal terms on the left-hand side and those on the right-hand side.

One hundred years after the appearance of Cayley's paper, Levine [61] generalized the identity (3.4) to 4×4 matrices. A year later Carlitz and Levine [63] produced the following generalization to $n\times n$ matrices.

THEOREM 3.4. *If $B=(b_{ij})$ is an n-square matrix of rank $\leqslant 2$ and without zero entries, then*

$$\operatorname{per}(B^{(-1)})\det(B^{(-1)})=\det(B^{(-2)}). \tag{3.7}$$

(Recall that $B^{(-1)}$ and $B^{(-2)}$ denote $n\times n$ matrices whose (i,j) entries are b_{ij}^{-1} and b_{ij}^{-2}, respectively.)

Proof. In Muir's identity (3.6), replace both a_{ij} and x_{ij} by b_{ij}^{-1}. Then the left-hand side of (3.6) becomes the left-hand side of (3.7), and the term on the right corresponding to the identity permutation is the right-hand side of (3.7). It remains to show that all the terms corresponding to other permutations vanish. Consider a permutation σ that is factored into a product of disjoint cycles. If any of the cycles is a transposition, then the corresponding matrix on the right-hand side of (3.6) has two identical rows, and thus its determinant vanishes.

Assume that the rows of B are linear combinations of the first two rows and that the first cycle of σ is $(1\,2\cdots h)$, where $3\leqslant h\leqslant n$.

Let $C=(c_{ij})$ be the $h\times n$ matrix consisting of the first h rows of the

ISBN 0-201-13505-1

matrix $B^{(-1)} * B_\sigma^{(-1)}$. Then

$$c_{ij} = b_{ij}^{-1} b_{i'j}^{-1},$$

where $i' = i+1$ for $i = 1,\ldots,h-1$, and $i' = 1$ if $i = h$. Let $K = \operatorname{diag}(k_1,\ldots,k_n)$, where

$$k_j = \prod_{s=1}^{h} b_{sj},$$

$j = 1,\ldots,n$. Then $D = (d_{ij}) = CK$ is an $h \times n$ matrix whose (i,j) entry is

$$d_{ij} = \prod_{\substack{s=1 \\ s \neq i,i'}}^{h} b_{sj},$$

$i = 1,\ldots,h$; $j = 1,\ldots,n$. Also, the rank of D is equal to that of C. But the rows of B are linear combinations of its first two rows; that is,

$$b_{sj} = \lambda_s b_{1j} + \mu_s b_{2j},$$

$s,j = 1,\ldots,n$. Thus,

$$\begin{aligned} d_{ij} &= \prod_{\substack{s=1 \\ s \neq i,i'}}^{h} (\lambda_s b_{1j} + \mu_s b_{2j}) \\ &= \sum_{t=0}^{h-2} \nu_{it} b_{1j}^{h-2-t} b_{2j}^{t}, \end{aligned}$$

so that the row space of D is spanned by $h-1$ vectors, and therefore the rank of C is less than h. It follows that the matrix corresponding to σ is singular. Hence the result. ■

Borchardt's identity (3.5) is an immediate corollary of Theorem 3.4, since the rank of $A^{(-1)}$ cannot exceed 2.

1.4 Renaissance of Permanents: Muirhead's Theorem, Pólya's Problem, Schur's Inequality and van der Waerden's Conjecture

The results on permanents of Borchardt, Cayley, Muir, and other writers in the nineteenth century consist of identities involving permanents and determinants. All these results are straightforward and elementary, although some of them are not easy to prove, owing to the complexity inherent in any multilinear function. The turning point in the history of

ISBN 0-201-13505-1

permanents came at the beginning of this century with the appearance of a beautiful theorem of Muirhead. The other three outstanding events that ushered the new era were: a problem posed by Pólya, the concept of generalized matrix functions introduced by Schur, and a conjecture proposed by van der Waerden.

If $\alpha=(\alpha_1,\ldots,\alpha_n)$ is a real n-tuple, let $\alpha^*=(\alpha_1^*,\ldots,\alpha_n^*)$ denote the n-tuple α rearranged in nonincreasing order, $\alpha_1^*\geqslant\cdots\geqslant\alpha_n^*$.

DEFINITION 4.1. A nonnegative n-tuple $\alpha=(\alpha_1,\ldots,\alpha_n)$ is said to be *majorized* by a nonnegative n-tuple $\beta=(\beta_1,\ldots,\beta_n)$, written $\alpha\prec\beta$, if

$$\alpha_1^*+\cdots+\alpha_k^*\leqslant\beta_1^*+\cdots+\beta_k^*$$

for $k=1,\cdots,n-1$, and

$$\alpha_1+\cdots+\alpha_n=\beta_1+\cdots+\beta_n.$$

THEOREM 4.1 (Muirhead [24]). *Let $c=(c_1,\ldots,c_n)$ be a positive n-tuple, and let $\alpha=(\alpha_1,\ldots,\alpha_n)$ and $\beta=(\beta_1,\ldots,\beta_n)$ be n-tuples of nonnegative integers. Let A and B be $n\times n$ matrices whose (i,j) entries are $c_i^{\alpha_j}$ and $c_i^{\beta_j}$, respectively. A necessary and sufficient condition that*

$$\operatorname{per}(A)\leqslant\operatorname{per}(B) \tag{4.1}$$

is that

$$\alpha\prec\beta. \tag{4.2}$$

Equality holds in (4.1) *if and only if either $\alpha=\beta$ or $c_1=\cdots=c_n$.*

Hardy, Littlewood, and Pólya [44] extended Muirhead's theorem to any nonnegative n-tuples α and β, proved the extended theorem, and showed that (4.2) is equivalent to the following condition: *There exists a doubly stochastic $n\times n$ matrix S such that*

$$\alpha=S\beta. \tag{4.3}$$

For proofs of Muirhead's theorem, see [44]. An outline of a proof that (4.2) and (4.3) are equivalent may also be found in [92].

Apart from its intrinsic interest, Muirhead's elegant result and its Hardy–Littlewood–Pólya generalization have had numerous important applications in many branches of mathematics.

The next landmark in the history of permanents was a question asked by Pólya [29]. As we shall see in the next chapters, most problems involving permanents present considerably more difficulty than the corresponding problems for determinants. It would therefore be of interest to find a

ISBN 0-201-13505-1

transformation that would convert permanents into determinants; specifically, given S, a set of $n\times n$ matrices, to find a linear transformation T on S such that

$$\operatorname{per}(T(A))=\det(A) \tag{4.4}$$

for all $A\in S$. Pólya's problem concerns transformations that involve only a uniform affixing of a plus or a minus sign to each position in the matrix. For example, if S is M_2, the set of 2×2 matrices, and

$$T\begin{bmatrix} a_{11} & a_{12} \\ a_{21} & a_{22} \end{bmatrix}=\begin{bmatrix} a_{11} & -a_{12} \\ a_{21} & a_{22} \end{bmatrix},$$

then (4.4) holds for all matrices in M_2. Similarly, if S is the set of all 3×3 matrices with a zero in the $(1,3)$ position, define T by

$$T\begin{bmatrix} a_{11} & a_{12} & 0 \\ a_{21} & a_{22} & a_{23} \\ a_{31} & a_{32} & a_{33} \end{bmatrix}=\begin{bmatrix} a_{11} & a_{12} & 0 \\ -a_{21} & a_{22} & a_{23} \\ a_{31} & -a_{32} & a_{33} \end{bmatrix}.$$

Then again $\operatorname{per}(T(A))=\det(A)$ for all A in S.

Pólya [29] asserted that, if S is the set of all $n\times n$ matrices, then for $n\geqslant 3$ there exists no transformation T involving uniform affixing of $\pm$ signs to entries of the matrices such that $\operatorname{per}(T(A))=\det(A)$.

The proof is quite simple. First, observe that it suffices to prove the assertion for $n=3$: If $n>3$, consider the direct sums of 3×3 matrices and the identity matrix I_{n-3}. Let $n=3$, and suppose that such a transformation exists. Then the number of minuses affixed to each diagonal corresponding to an even permutation would have to be even, and thus the total number of minuses in the transformation must be even. On the other hand, the number of minuses on each diagonal corresponding to an odd permutation would have to be odd, and hence the total number of minuses would have to be a sum of three odd numbers—that is, an odd number. Contradiction.

Pólya's result was substantially generalized by Marcus and Minc [70], who showed that for $n\geqslant 3$ there exists no linear transformation T on the set of $n\times n$ matrices such that $\operatorname{per}(T(A))=\det(A)$. This result finally and completely closed the door on any hope that problems involving permanents can be solved easily via determinants.

As we saw in one of our examples above, in certain sets of matrices the permanent can be converted into the determinant by affixing in a uniform way $\pm$ signs to the matrix entries. In fact, Gibson [192] showed that, if an n-square $(0,1)$-matrix A has a positive permanent and if the permanent of A can be converted into a determinant by affixing $\pm$ signs to the entries of A, then A has at most $(n^2+3n-2)/2$ ones.

ISBN 0-201-13505-1

An entirely new approach to permanents and determinants was initiated in 1918 by Schur [34], who introduced the concept of *generalized matrix functions* (often called *Schur functions*) on square matrices. Let $A=(a_{ij})\in M_n(\mathbf{C})$, the set of $n\times n$ complex matrices (or matrices over any commutative ring with 1). Let H be a subgroup of the symmetric group S_n, and let $\chi: H\to\mathbf{C}$ be a scalar-valued function on H. We define a function $d_\chi^H: M_n(\mathbf{C})\to\mathbf{C}$ by

$$d_\chi^H(A)=\sum_{\sigma\in H}\chi(\sigma)\prod_{i=1}^{n}a_{i\sigma(i)}. \tag{4.5}$$

Of particular interest is the case when χ is a character of degree 1—that is, a nontrivial homomorphism from H to $\mathbf{C}$. We shall call d_χ^H a *Schur function* if this is the case. In particular, if $H=S_n$ and $\chi=\varepsilon$ (that is, χ is the sign function), then d_χ^H is simply the determinant function. If $H=S_n$ and $\chi\equiv 1$, then it is the permanent function.

The study of Schur functions has made considerable progress in the past decade. Schur himself obtained the following elegant result, which is given here in its specialization to permanents.

THEOREM 4.2 (Schur [34]). *If A is a positive semi-definite hermitian matrix, then*

$$\det(A)\leqslant \operatorname{per}(A). \tag{4.6}$$

Equality can hold in (4.6) *if and only if A is diagonal or A has a zero row.*

We shall prove this result in the next chapter.

The main cause of the current great revival of interest in the theory of permanents is the so-called van der Waerden conjecture. It started as an innocent problem proposed by van der Waerden [37]: What is the minimum value of the permanent function on Ω_n, the set of doubly stochastic $n\times n$ matrices (that is, nonnegative matrices with row sums and column sums equal to 1)? The conjectured answer to this question is known as the *van der Waerden conjecture*. Let J_n denote the matrix all of whose entries equal $1/n$. If $S\in\Omega_n$ and $S\neq J_n$, then it is conjectured that

$$\begin{aligned}\operatorname{per}(S)&>\operatorname{per}(J_n)\\ &=n!/n^n.\end{aligned}$$

The conjecture is still unresolved. We shall discuss its current status in Chapter 5.

ISBN 0-201-13505-1

1.5 The New Era: Marvin Marcus and Company

In the period between 1926 and 1959, the theory of permanents remained dormant. It is true that many of the combinatorial results obtained during this period could be expressed in terms of permanents, but the authors invariably used combinatorial methods without even mentioning permanents.

A dramatic reawakening of interest occurred in 1959 when, almost simultaneously, Brenner [58] showed that permanents of matrices with a dominant diagonal cannot vanish, Caianiello [59, 60] made use of permanents to express the perturbation expansions of quantum field theory in a compact algebraic form, Levine [61] extended Cayley's identity (Theorem 3.1) to 4×4 matrices, and Marcus and Newman [62] published their historical paper on the van der Waerden conjecture. Indeed, the last paper was the *causa causans* of the present remarkable activity in the area of permanent theory that has prompted over 100 authors to contribute over 200 papers to this field of mathematics. A substantial part of this work was done by Marcus, his associates, students, and former students.

In the area of the van der Waerden conjecture, apart from the first Marcus–Newman paper [62], the contents of which we shall discuss in Chapter 5, Marcus and Newman [79] proved the conjecture for positive semi-definite symmetric matrices. Sasser and Slater [134] extended this theorem to normal matrices of a certain type. The Sasser–Slater result was then somewhat improved by Marcus and Minc [144] and extended to a larger class of matrices by Friedland [243].

Marcus and Newman [62] also proved the van der Waerden conjecture for all doubly stochastic 3×3 matrices. Eberlein and Mudholkar [139] succeeded in proving it for 4×4 matrices. This case was re-proved by Gleason [177]. Then Eberlein [152] proved the conjecture for 5×5 matrices, and there the matter rests at present.

As for Ω_n, the set of all doubly stochastic $n \times n$ matrices, Marcus and Minc [77] proved that the permanent of any matrix in Ω_n is at least n^{-n}. Rothaus [253] showed that the bound can be improved to n^{1-n}, and Friedland [296] obtained a substantially better bound of $1/n!$. Recently Bang [274] announced, without proof, a bound that is essentially of the same order as the van der Waerden bound.

The theory of permanents has made solid progress on many other fronts. On the subpermanent front the main interest has been in the function $\sigma_t(A)$, the sum of all subpermanents of A of order t. Tverberg [88] showed that, for a doubly stochastic $n \times n$ matrix $A, \sigma_t(A) \geqslant \sigma_t(J_n)$ for $t=2$ and 3, where J_n denotes the $n \times n$ matrix all of whose entries equal $1/n$. Đoković [123] conjectured a lower bound for the ratio $\sigma_t(A)/\sigma_{t-1}(A)$ for a doubly stochastic A, and proved it for $t=3$. Marcus and Minc [144] proved the

ISBN 0-201-13505-1

conjecture for any t but only for certain normal doubly stochastic matrices (see Section 8.1).

As we mentioned in Section 1.3, Marcus and Minc [70] showed that there exists no linear transformation T on M_n, the set of $n \times n$ matrices, such that $\operatorname{per}(T(A)) = \det(A)$. Marcus and May [76] determined the form of linear transformations that keep the permanent fixed. Both results were re-proved by Botta [122, 137], and an interesting postscript to the Marcus–Minc result was added by Gibson [192]. Minc [279] obtained the form of linear transformations on M_n that hold the permanent fixed and preserve either the trace or the second elementary symmetric function of the eigenvalues.

Since 1935 the permanents of (0,1)-matrices have made their appearance in many combinatorial investigations but often under some other name. The front of permanents of (0,1)-matrices was opened in 1960 by Ryser [67,68], Nikolai [66], and Tinsley [69], who studied mainly the permanents of incidence matrices of combinatorial configurations. Mendelsohn [73], Minc [94], and Metropolis, Stein,and Stein [160] obtained formulas for the permanents of (0,1)-circulants. In 1963 Minc [84] conjectured an upper bound for the permament of a (0,1)-matrix in terms of its row sums, and proved a weaker version of the conjecture. Jurkat and Ryser [112] used their remarkable expression for the permanent as a product of matrices to improve Minc's upper bound. Other improvements were obtained by Minc [132] and by Nijenhuis and Wilf [182, 183]. Finally, Brégman [219] succeeded in proving the conjecture.

Sinkhorn and Knopp [170] obtained a canonical form for nearly decomposable matrices. This enabled Sinkhorn [169], Minc [162], Gibson [206], Hartfiel [229], and others, to obtain lower bounds for the permanents of (0,1)-matrices.

Many interesting inequalities for the permanent function have been obtained by Marcus [89, 100], Marcus and Gordon [91], Marcus and Minc [93, 101], Perfect [95], Brualdi and Newman [97], Brualdi [107], Lieb [114], Baum and Eagon [120], Brenner and Brualdi [121], Đoković [124], and others. We shall discuss some of these results in subsequent chapters.

One of the most important developments in the theory of permanents has been the introduction of the methods of multilinear algebra to the study of Schur functions. In fact, the van der Waerden conjecture for positive semi-definite matrices [79], the Hadamard theorem for permanents [89], an upper bound for the permanents of normal matrices [93], and many other inequalities have been obtained by these methods. Multilinear algebra, and even the general theory of Schur functions, is beyond the scope of this book. We shall introduce in the next chapter enough of multilinear algebra to present some samples of the use of these techniques in the theory of permanents.

ISBN 0-201-13505-1

Problems

1. Show that, if A is a nonnegative $n\times n$ matrix, then

$$|\det(A)| \leqslant \operatorname{per}(A) \leqslant \prod_{i=1}^{n} r_i.$$

Is either inequality true for all real $n\times n$ matrices?
2. Let

$$A=\begin{bmatrix} 1 & 2 & 0 & 1 \\ 2 & -1 & -1 & 3 \\ 1 & 2 & 1 & 0 \\ 0 & 1 & 3 & 1 \end{bmatrix}.$$

Use Binet's formula to evaluate $\operatorname{per}(A)$.
3. Let $t_1,\dots,t_k$ be integers, $1 \leqslant t_1 \leqslant \cdots \leqslant t_k$, $t_1+\cdots+t_k=m$. Let the multiplicity of t_i be $m_i, i=1,\dots,k, m_1+\cdots+m_k=k$. Show that the number of terms in $S(t_1,\dots,t_k)$ is

$$m!/\prod_{i=1}^{k} t_i!(m_i!)^{1/m_i}.$$

4. Use direct combinatorial reasoning to obtain a Binet-type formula for the permanents of $5\times n$ matrices.
5. Use Muirhead's theorem to prove that the geometric mean of nonnegative numbers cannot exceed their arithmetic mean. Deduce the condition for equality.
6. Show that the following generalization of Gibson's theorem in [192] (see Section 1.4) is invalid: Let A be an n-square real matrix with nonzero permanent. If the permanent of A can be converted into a determinant by affixing $\pm$ signs to the entries of A, then A has at most $(n^2+3n-2)/2$ nonzero entries.
7. Show that the proviso in Gibson's theorem in [192], that $\operatorname{per}(A)>0$, is essential.
8. Show that the converse of Gibson's theorem in [192] is not true.
9. Prove that, if χ is a character on H, and $H=S_n$, then either $\chi=\varepsilon$ or $\chi\equiv 1$. In other words, show that the only Schur functions involving all the permutations in S_n are the determinant and the permanent.

ISBN 0-201-13505-1

CHAPTER 2

Properties of Permanents

2.1 Elementary Properties

Owing to the similarity in the definitions of permanents and determinants, many of the properties of determinants have their analogues for permanents. In some cases, as in the first theorems below, the properties of permanents are easier to establish than those of determinants. Unfortunately, the permanents fail to inherit two key properties of determinants: the multiplicative property and the invariance under certain elementary operations on matrices. It is this deficiency that accounts for the fact that proving theorems on permanents, and evaluating them, is substantially more difficult, in general, than solving analogous problems for determinants.

We shall require the following simplifying notation. Let $\Gamma_{r,n}$ denote the set of all n^r sequences $\omega=(\omega_1,\ldots,\omega_r)$ of integers, $1 \leqslant \omega_i \leqslant n$, $i=1,\ldots,n$. Let $G_{r,n}$ denote the subset of $\Gamma_{r,n}$ consisting of all nondecreasing sequences,

$$G_{r,n}=\{(\omega_1,\ldots,\omega_r)\in\Gamma_{r,n}|1 \leqslant \omega_1 \leqslant \cdots \leqslant \omega_r \leqslant n\},$$

and let $Q_{r,n}$ be the set of increasing sequences,

$$Q_{r,n}=\{(\omega_1,\ldots,\omega_r)\in\Gamma_{r,n}|1 \leqslant \omega_1 < \cdots < \omega_r \leqslant n\}.$$

For a sequence $\omega\in\Gamma_{r,n}$, let $m_t(\omega)$ denote the number of occurrences of t in ω. For example, $m_1((2,4,4,4,5,5))=0$; $m_2((2,4,4,4,5,5))=1$; $m_3((2,4,4,4,5,5))=0$; $m_4((2,4,4,4,5,5))=3$; and $m_5((2,4,4,4,5,5))=2$. Next, let $\mu(\omega)$ denote the product of the factorials of the multiplicities of

ENCYCLOPEDIA OF MATHEMATICS and Its Applications, Gian-Carlo Rota (ed.). Vol. 6: Henryk Minc, Permanents

ISBN 0-201-13505-1

the distinct integers in ω; that is,

$$\mu(\omega)=\prod_{t=1}^{n} m_t(\omega)!.$$

For example, $\mu((2,4,4,4,5,5))=3!2!=12$.

We shall also require the following matrix notation (see [92]). Let $M_{m,n}(S)$, or simply $M_{m,n}$, denote the set of all $m\times n$ matrices with entries from a set S. If $m=n$, we shall write M_n instead of $M_{n,n}$. Now, let $A=(a_{ij})\in M_{m,n}$, and let $\alpha\in G_{h,m}$ and $\beta\in G_{k,n}$. Then $A[\alpha|\beta]$ denotes the $h\times k$ matrix whose (i,j) entry is $a_{\alpha_i\beta_j}$. If it happens that $\alpha\in Q_{h,m}$ and $\beta\in Q_{k,n}$, then $A[\alpha|\beta]$ is a submatrix of A. If $\alpha=\beta$, we simplify the notation to $A[\alpha]$. Again, if $\alpha\in Q_{h,m}$ and $\beta\in Q_{k,n}$, then $A(\alpha|\beta)$ denotes the $(m-h)\times(n-k)$ submatrix of A complementary to $A[\alpha|\beta]$—that is, the submatrix obtained from A by deleting rows α and columns β. In particular, the $(m-h)\times n$ submatrix obtained from A by deleting rows α is denoted by $A(\alpha|-)$. Similarly, $A(-|\beta)$ denotes the $m\times(n-k)$ submatrix obtained from A by deleting columns β.

THEOREM 1.1.(a) *The permanent function on $m\times n$ matrices, $m\leqslant n$, is a multilinear function of the rows of each matrix. If $m=n$, it is also a multilinear function of the columns.*

(b) *If A is an $m\times n$ matrix, $m\leqslant n$, and P and Q are permutation matrices of orders m and n, respectively, then*

$$\operatorname{Per}(PAQ)=\operatorname{Per}(A).$$

(c) *If A is an n-square matrix, then*

$$\operatorname{per}(A^T)=\operatorname{per}(A).$$

All these properties are immediate consequences of the definition of permanents.

Our next theorem is an analogue of the Laplace expansion theorem for determinants.

THEOREM 1.2. *If A is an $m\times n$ matrix, $2\leqslant m\leqslant n$, and $\alpha\in Q_{r,m}$, then*

$$\operatorname{Per}(A)=\sum_{\omega\in Q_{r,m}}\operatorname{Per}(A[\alpha|\omega])\operatorname{Per}(A(\alpha|\omega)). \tag{1.1}$$

In particular, for any i, $1\leqslant i\leqslant m$,

$$\operatorname{Per}(A)=\sum_{t=1}^{n} a_{it}\operatorname{Per}(A(i|t)). \tag{1.2}$$

ISBN 0-201-13505-1

If $m=n$ and $\beta\in Q_{r,n}$, then also

$$\operatorname{per}(A)=\sum_{\omega\in Q_{r,n}}\operatorname{per}(A[\omega|\beta])\operatorname{per}(A(\omega|\beta)), \tag{1.3}$$

and for any j, $1\leqslant j\leqslant n$,

$$\operatorname{per}(A)=\sum_{t=1}^{n}a_{tj}\operatorname{per}(A(t|j)). \tag{1.4}$$

The proof of (1.1) is quite straightforward. In fact, it is much easier than the corresponding theorem for determinants. We leave it as an exercise for the reader (Problem 2). Formula (1.3) is a consequence of Theorem 1.1(c).

Before we state our next theorem, which is an analogue of the Binet–Cauchy theorem for determinants, we prove a combinatorial lemma.

LEMMA 1. *Let f be a scalar function on m-tuples of integers. Then*

$$\sum_{\omega\in\Gamma_{m,n}}f(\omega_1,\ldots,\omega_m)=\sum_{\omega\in G_{m,n}}\frac{1}{\mu(\omega)}\sum_{\sigma\in S_m}f(\omega_{\sigma 1},\ldots,\omega_{\sigma m}), \tag{1.5}$$

where $\omega=(\omega_1,\ldots,\omega_m)$.

Proof. Partition the set of m-tuples $\Gamma_{m,n}$ into equivalence classes: Two m-tuples are equivalent if they contain exactly the same integers. Then each class contains exactly one m-tuple belonging to $G_{m,n}$. If we permute the integers $\omega_1,\ldots,\omega_m$ of an m-tuple $\omega=(\omega_1,\ldots,\omega_m)\in G_{m,n}$ in all possible $m!$ ways, we obtain all the m-tuples in the equivalence class of ω, each m-tuple appearing exactly $\mu(\omega)$ times. We use this observation to sum up the $f(\omega_1,\ldots,\omega_m)$ on the right-hand side of (1.5). This is accomplished by summing up the $f(\omega_1,\ldots,\omega_m)$ separately in each equivalence class and then adding all these partial sums. ■

THEOREM 1.3 (Binet–Cauchy theorem for permanents). *If B and C are $m\times n$ and $n\times m$ matrices, respectively, $m\leqslant n$, then*

$$\operatorname{per}(BC)=\sum_{\omega\in G_{m,n}}\frac{1}{\mu(\omega)}\operatorname{per}(B[1,\ldots,m|\omega])\operatorname{per}(C[\omega|1,\ldots,m]).$$

Proof. It is convenient to regard the permanent function as a function on n-tuples of vectors. If $A_{(1)},\ldots,A_{(m)}$ denote the rows of an $m\times m$ matrix A, we define

$$\operatorname{per}(A_{(1)},\ldots,A_{(m)})=\operatorname{per}(A).$$

ISBN 0-201-13505-1

Now,

$$(BC)_{(i)} = \sum_{t=1}^{n} b_{it} C_{(t)},$$

and therefore, by Theorem 1.1(a),

$$\begin{aligned}
\operatorname{per}(BC) &= \operatorname{per}((BC)_{(1)}, \ldots, (BC)_{(m)}) \\
&= \operatorname{per}\left(\sum_{t=1}^{n} b_{1t} C_{(t)}, \ldots, \sum_{t=1}^{n} b_{mt} C_{(t)} \right) \\
&= \sum_{\omega \in \Gamma_{m,n}} \prod_{i=1}^{m} b_{i\omega_i} \operatorname{per}(C_{(\omega_1)}, \ldots, C_{(\omega_m)}) \\
&= \sum_{\omega \in G_{m,n}} \frac{1}{\mu(\omega)} \sum_{\sigma \in S_m} \prod_{i=1}^{m} b_{i\omega_{\sigma i}} \operatorname{per}(C_{(\omega_{\sigma 1})}, \ldots, C_{(\omega_{\sigma m})}),
\end{aligned}$$

by the lemma. But by Theorem 1.1(b),

$$\begin{aligned}
\operatorname{per}(C_{(\omega_{\sigma 1})}, \ldots, C_{(\omega_{\sigma m})}) &= \operatorname{per}(C_{(\omega_1)}, \ldots, C_{(\omega_m)}) \\
&= \operatorname{per}(C[\omega|1, \ldots, m]),
\end{aligned}$$

and therefore

$$\begin{aligned}
\operatorname{per}(BC) &= \sum_{\omega \in G_{m,n}} \frac{1}{\mu(\omega)} \operatorname{per}(C[\omega|1, \ldots, m]) \sum_{\sigma \in S_m} \prod_{i=1}^{m} b_{i\omega_{\sigma i}} \\
&= \sum_{\omega \in G_{m,n}} \frac{1}{\mu(\omega)} \operatorname{per}(B[1, \ldots, m|\omega]) \operatorname{per}(C[\omega|1, \ldots, m]). \qquad \blacksquare
\end{aligned}$$

Our next permanental identity is less known than the preceding results. It seems to have been noted first by Caianiello [59] as an analogue to an identity for determinants apparently due to Arnaldi (see also [144] and [156]).

THEOREM 1.4. *Let $A = (a_{ij})$ and $B = (b_{ij})$ be $n \times n$ matrices. Then*

$$\operatorname{per}(A + B) = \sum_{r=0}^{n} \sum_{\alpha, \beta \in Q_{r,n}} \operatorname{per}(A[\alpha|\beta]) \operatorname{per}(B(\alpha|\beta)),$$

where $\operatorname{per}(A[\alpha|\beta]) = 1$ *and* $\det(B(\alpha|\beta)) = \det(B)$ *when* $r = 0$, *and* $\det(B(\alpha|\beta)) = 1$ *when* $r = n$.

The following proof was suggested by Marvin Marcus. First, note the

ISBN 0-201-13505-1

following identity:

$$\prod_{i=1}^{n}(x_i+y_i)=\sum_{r=0}^{n}\sum_{\alpha\in Q_{r,n}}\prod_{i=1}^{n}x_{\alpha_i}\prod_{i=r+1}^{n}y_{\alpha_i'} \tag{1.6}$$

where $\alpha=(\alpha_1,\dots,\alpha_r)\in Q_{r,n}$, and $\alpha'=(\alpha_{r+1}',\dots,\alpha_n')\in Q_{n-r,n}$ is the sequence complementary to α in $(1,\dots,n)$. Hence

$$\begin{aligned}\operatorname{per}(A+B)&=\sum_{\sigma\in S_n}\prod_{i=1}^{n}(a_{i,\sigma i}+b_{i,\sigma i})\\&=\sum_{\sigma\in S_n}\sum_{r=0}^{n}\sum_{\alpha\in Q_{r,n}}\prod_{i=1}^{r}a_{\alpha_i,\sigma\alpha_i}\prod_{i=r+1}^{n}b_{\alpha_i',\sigma\alpha_i'}\\&=\sum_{r=0}^{n}\sum_{\alpha\in Q_{r,n}}\left(\sum_{\sigma\in S_n}\prod_{i=1}^{r}a_{\alpha_i,\sigma\alpha_i}\prod_{i=r+1}^{n}b_{\alpha_i',\sigma\alpha_i'}\right).\end{aligned}$$

Now, the expression in the parentheses is the permanent of the matrix whose ith row is the (α_i)th row of A for $i=1,\dots,r$, and the (α_i')th row of B for $i=r+1,\dots,n$. But by Theorem 1.2, expanding by rows indexed α, the permanent of the matrix is

$$\sum_{\beta\in Q_{r,n}}\operatorname{per}(A[\alpha|\beta])\operatorname{per}(B(\alpha|\beta)).$$

The result follows. ■

2.2 The Permanent Function as an Inner Product

A major breakthrough in the theory of inequalities involving permanents occurred in 1961, when Marcus and Newman [72, 79] developed a new approach to the permanent function by representing it as an inner product on the symmetry class of completely symmetric tensors and applying to it the Cauchy–Schwarz inequality. We shall not attempt here to give a full exposition of the theory of tensor spaces; this can be found in many standard works (see, e.g., [233]). However, for the purpose of reference we shall discuss some of the underlying ideas.

Let V be an n-dimensional unitary space with inner product (u,v). Let $M_m(V)$ be the space of m-multilinear functionals on V. In other words, $M_m(V)$ is the space of complex-valued functions φ on $V\times V\times\cdots\times V$, the set of m-tuples of vectors in V, satisfying:

$$\begin{aligned}\varphi(u_1,\dots,u_{i-1},cu_i+c'u_i',u_{i+1},\dots,u_m)&=c\varphi(u_1,\dots,u_{i-1},u_i,u_{i+1},\dots,u_m)\\&\quad+c'\varphi(u_1,\dots,u_{i-1},u_i',u_{i+1},\dots,u_m),\end{aligned}$$

ISBN 0-201-13505-1

for any $u_1,\ldots,u_i,u_i',\ldots,u_m$ in V, any i, and any complex numbers c and c'. For example, if V is $V_n(\mathbf{C})$, the space of complex n-tuples, and $u_i=(a_{i1},\ldots,a_{in})$, $i=1,\ldots,m$, then the multilinear functional φ defined by

$$\varphi(u_1,\ldots,u_m)=\operatorname{Per}((a_{ij}))$$

is in $M_m(V)$. If $m=n$, $V=V_n(\mathbf{C})$, and ψ is defined by

$$\psi(u_1,\ldots,u_n)=\det((a_{ij}))$$

for all n-tuples $u_1,\ldots,u_n$, then ψ is in $M_n(V)$.

Definition 2.1. If V is an n-dimensional unitary space, then the dual space of $M_m(V)$ is called the *space of m-contravariant tensors* (or simply *tensors*) and is denoted by $V^{(m)}$. In other words, $V^{(m)}$ is the space of all complex-valued linear functionals on $M_m(V)$, the space of m-multilinear functionals on V. If $u_i\in V$, $i=1,\ldots,m$, then we define $f=u_1\otimes\cdots\otimes u_m$ as the element of $V^{(m)}$ whose value on any $\varphi\in M_m(V)$ is given by

$$f(\varphi)=\varphi(u_1,\ldots,u_m).$$

The tensor $u_1\otimes\cdots\otimes u_m$ is said to be *decomposable*. It is also called the *tensor product* of the $u_1,\ldots,u_m$.

Not all tensors of $V^{(m)}$ are decomposable, but they all are linear combinations of decomposable tensors. In fact, it is easy to show that, if $e_1,\ldots,e_n$ is a basis of V, then the n^m tensor products

$$e_{\omega_1}\otimes\cdots\otimes e_{\omega_m}$$

(where the ω_i run independently over the integers $1,\ldots,n$) constitute a basis of $V^{(m)}$.

We now introduce an inner product in $V^{(m)}$ induced by the inner product in V. We define it in terms of its values on pairs of decomposable tensors:

$$(u_1\otimes\cdots\otimes u_m, v_1\otimes\cdots\otimes v_m)=\prod_{i=1}^{m}(u_i,v_i). \tag{2.1}$$

Example 2.1. Let A and B be $m\times m$ matrices. If we set $u_i=A_{(i)}$, $v_i=\overline{B}^{(i)}$, $i=1,\ldots,m$, (where $B^{(i)}$ denotes the ith column of B) in (2.1), we obtain

$$\begin{aligned}\left(A_{(1)}\otimes\cdots\otimes A_{(m)},\overline{B}^{(1)}\otimes\cdots\otimes\overline{B}^{(m)}\right)&=\prod_{i=1}^{m}\left(A_{(i)},\overline{B}^{(i)}\right)\\&=\prod_{i=1}^{m}(AB)_{ii}.\end{aligned} \tag{2.2}$$

ISBN 0-201-13505-1

Applying the Cauchy–Schwarz inequality to (2.2), we obtain

$$\prod_{i=1}^{m} |(AB)_{ii}|^2 \leqslant \prod_{i=1}^{m} (AA^*)_{ii} \prod_{i=1}^{m} (B^*B)_{ii}$$
$$= \prod_{i=1}^{m} \|A_{(i)}\|^2 \|B^{(i)}\|^2.$$

DEFINITION 2.2. Let σ be a permutation in S_m. Define a linear operator $P(\sigma)$ on $V^{(m)}$ by

$$P(\sigma)(u_1 \otimes \cdots \otimes u_m) = u_{\sigma^{-1}(1)} \otimes \cdots \otimes u_{\sigma^{-1}(m)}. \tag{2.3}$$

The *completely symmetric operator* on $V^{(m)}$ is now defined by

$$T_m = \frac{1}{m!} \sum_{\sigma \in S_m} P(\sigma). \tag{2.4}$$

The range of T_m is denoted by $V_{(m)}$ and is called the *symmetry class* of completely symmetric tensors. We also introduce the following notation:

$$u_1 * \cdots * u_m = T_m(u_1 \otimes \cdots \otimes u_m).$$

The tensor $u_1 * \cdots * u_m$ is called the *symmetric product* of the $u_1, \ldots, u_m$.

THEOREM 2.1. *If $u_i \in V$, $i = 1, \ldots, m$, then:*
(a) $P(\sigma)(u_1 * \cdots * u_m) = u_1 * \cdots * u_m$;
(b) $T_m^2 = T_m$;
(c) $T_m^* = T_m$;
that is, T_m is hermitian with respect to the inner product (2.1).

Proof. (a) We have

$$\begin{aligned} P(\sigma)u_1 * \cdots * u_m &= P(\sigma)T_m(u_1 \otimes \cdots \otimes u_m) \\ &= \frac{1}{m!} P(\sigma) \sum_{\varphi \in S_m} P(\varphi)(u_1 \otimes \cdots \otimes u_m) \\ &= \frac{1}{m!} P(\sigma) \sum_{\varphi \in S_m} u_{\varphi^{-1}(1)} \otimes \cdots \otimes u_{\varphi^{-1}(m)} \\ &= \frac{1}{m!} \sum_{\varphi \in S_m} u_{\sigma^{-1}\varphi^{-1}(1)} \otimes \cdots \otimes u_{\sigma^{-1}\varphi^{-1}(m)} \\ &= \frac{1}{m!} \sum_{\tau \in S_m} u_{\tau^{-1}(1)} \otimes \cdots \otimes u_{\tau^{-1}(m)} \\ &= \frac{1}{m!} \sum_{\tau \in S_m} P(\tau)(u_1 \otimes \cdots \otimes u_m) \\ &= u_1 * \cdots * u_m. \end{aligned}$$

ISBN 0-201-13505-1

(b) It suffices to prove that T_m is idempotent when operating on decomposable tensors. If $u_1,\dots,u_m$ are any vectors in V, then

$$\begin{aligned}T_m^2(u_1\otimes\cdots\otimes u_m)&=T_m(u_1*\cdots*u_m)\\&=\frac{1}{m!}\sum_{\sigma\in S_m}P(\sigma)(u_1*\cdots*u_m)\\&=\frac{1}{m!}\sum_{\sigma\in S_m}u_1*\cdots*u_m\\&=u_1*\cdots*u_m\\&=T_m(u_1\otimes\cdots\otimes u_m).\end{aligned}$$

(c) Again it suffices to prove the hermitian property for decomposable tensors:

$$\begin{aligned}&(T_m(u_1\otimes\cdots\otimes u_m),v_1\otimes\cdots\otimes v_m)\\&=\frac{1}{m!}\sum_{\sigma\in S_m}(u_{\sigma^{-1}(1)}\otimes\cdots\otimes u_{\sigma^{-1}(m)},v_1\otimes\cdots\otimes v_m)\\&=\frac{1}{m!}\sum_{\sigma\in S_m}\prod_{i=1}^{n}(u_{\sigma^{-1}(i)},v_i)\\&=\frac{1}{m!}\sum_{\sigma\in S_m}\prod_{i=1}^{n}(u_i,v_{\sigma(i)})\\&=\frac{1}{m!}\sum_{\sigma^{-1}\in S_m}(u_1\otimes\cdots\otimes u_m,v_{\sigma(1)}\otimes\cdots\otimes v_{\sigma(m)})\\&=(u_1\otimes\cdots\otimes u_m,T_m(v_1\otimes\cdots\otimes v_m)).\end{aligned}$$

■

THEOREM 2.2. *Let $u_1,\dots,u_m$, $v_1,\dots,v_m$ be vectors in V, and let $A=(a_{ij})=((u_i,v_j))$. Then*

$$\begin{aligned}(u_1*\cdots*u_m,v_1*\cdots*v_m)&=\frac{1}{m!}\operatorname{per}((u_i,v_j))\\&=\frac{1}{m!}\operatorname{per}(A).\end{aligned}\tag{2.5}$$

Proof. We use Theorem 2.1 and formula (2.1) to compute

$$\begin{aligned}(u_1*\cdots*u_m, v_1*\cdots*v_m) &= (T_m(u_1\otimes\cdots\otimes u_m), T_m(v_1\otimes\cdots\otimes v_m))\\ &= (T_m^*T_m(u_1\otimes\cdots\otimes u_m, v_1\otimes\cdots\otimes v_m))\\ &= (T_m(u_1\otimes\cdots\otimes u_m, v_1\otimes\cdots\otimes v_m))\\ &= \frac{1}{m!}\sum_{\sigma\in S_m}(P(\sigma)u_1\otimes\cdots\otimes u_m, v_1\otimes\cdots\otimes v_m)\\ &= \frac{1}{m!}\sum_{\sigma\in S_m}(u_{\sigma^{-1}(1)}\otimes\cdots\otimes u_{\sigma^{-1}(m)}, v_1\otimes\cdots\otimes v_m)\\ &= \frac{1}{m!}\sum_{\sigma\in S_m}\prod_{i=1}^m(u_{\sigma^{-1}(i)}, v_i)\\ &= \frac{1}{m!}\sum_{\sigma\in S_m}\prod_{i=1}^m(u_i, v_{\sigma(i)})\\ &= \frac{1}{m!}\sum_{\sigma\in S_m}\prod_{i=1}^m a_{i\sigma(i)}\\ &= \frac{1}{m!}\operatorname{per}(A).\end{aligned}$$

■

THEOREM 2.3 [79].
(a) *The symmetric product* $u_1*\cdots*u_m$ *is* 0 *if and only if* $u_i=0$ *for some* i.
(b) *Let* $u_1,\ldots,u_m,v_1,\ldots,v_m$ *be nonzero vectors in* V. *Then*

$$u_1*\cdots*u_m=v_1*\cdots*v_m$$

if and only if

$$v_i=d_iu_{\sigma(i)},\qquad i=1,\ldots,m,$$

where σ *is a permutation in* S_m, *and* $d_1,\ldots,d_m$ *are scalars,* $d_1\cdots d_m=1$.

Proof. (a) Suppose that $u_1*\cdots*u_m=0$; that is, $T_m(u_1\otimes\cdots\otimes u_m)=0$. Now, for an arbitrary vector $w\in V$,

$$\begin{aligned}(T(u_1\otimes\cdots\otimes u_m), w\otimes\cdots\otimes w) &= (u_1\otimes\cdots\otimes u_m, T(w\otimes\cdots\otimes w))\\ &= (u_1\otimes\cdots\otimes u_m, w\otimes\cdots\otimes w)\\ &= \prod_{i=1}^m(u_i,w).\end{aligned}\tag{2.6}$$

ISBN 0-201-13505-1

Hence

$$\prod_{i=1}^{m} (u_i, w) = 0 \tag{2.7}$$

for all $w \in W$. Thus u_i must be 0 for some i. For, if $u_i \neq 0$, $i = 1, \ldots, m$, then $\dim\langle u_i\rangle^\perp = n-1$, $i = 1, \ldots, m$, and thus

$$\bigcup_{i=1}^{m} \langle u_i\rangle^\perp \neq V;$$

then for any $w \notin \bigcup_{i=1}^{m}\langle u_i\rangle^\perp$, the product (2.7) would not be 0.

(b) If $u_1 * \cdots * u_m = v_1 * \cdots * v_m$, then by (2.6), for any $w \in V$,

$$\prod_{i=1}^{m} (u_i, w) = \prod_{i=1}^{m} (v_i, w). \tag{2.8}$$

Now let w be any vector in $\langle u_1\rangle^\perp$, and let $v_i = x_i + y_i$, $x_i \in \langle u_1\rangle$, $y_i \in \langle u_1\rangle^\perp$. Then

$$0 = \prod_{i=1}^{m} (v_i, w) = \prod_{i=1}^{m} (y_i, w).$$

Hence some y_i, say y_1, must be 0. But then $v_1 = x_1 = d_1 u_1$. Thus (2.8) becomes

$$(u_1, w)\left(\prod_{i=2}^{m} (u_i, w) - d_1 \prod_{i=2}^{m} (v_i, w) \right) = 0$$

for all $w \in V$. Since $u_1 \neq 0$, this implies that

$$\begin{aligned} \prod_{i=2}^{m} (u_i, w) &= d_1 \prod_{i=2}^{m} (v_i, w) \\ &= (d_1 v_2, w)(v_3, w) \cdots (v_m, w), \end{aligned}$$

for all $w \in V$, and induction on m completes the proof. ∎

We are now ready for the fundamental inequality of Marcus and Newman [79]. We shall state it in two essentially equivalent forms.

THEOREM 2.4 (Marcus–Newman).

(a) *If $u_1, \ldots, u_m, v_1, \ldots, v_m$ are vectors in a unitary space V, then*

$$\left|\operatorname{per}((u_i, v_j))\right|^2 \leqslant \operatorname{per}((u_i, u_j)) \operatorname{per}((v_i, v_j)). \tag{2.9}$$

ISBN 0-201-13505-1

Equality holds in (2.9) *if and only if either one of the* $u_1,\ldots,u_m$, $v_1,\ldots,v_m$ *is zero, or there exists a permutation* σ *such that* $u_i = d_i v_{\sigma(i)}$, $i=1,\ldots,m$, *for some scalars* $d_1,\ldots,d_m$.

(b) *If* A *is an* $m\times n$ *matrix and* B *is an* $n\times m$ *matrix, then*

$$|\operatorname{per}(AB)|^2 \leqslant \operatorname{per}(AA^*)\operatorname{per}(B^*B). \tag{2.10}$$

Equality holds in (2.10) *if and only if either* A *has a zero row, or* B *has a zero column, or* $A = DPB^*$ *for some diagonal matrix* D *and a permutation matrix* P.

Proof. (a) From the Cauchy–Schwarz inequality we have

$$|(u_1*\cdots*u_m, v_1*\cdots*v_m)|^2 \leqslant \|u_1*\cdots*u_m\|^2\|v_1*\cdots*v_m\|^2, \tag{2.11}$$

and by Theorem 2.2,

$$|(u_1*\cdots*u_m, v_1*\cdots*v_m)|^2 = \frac{1}{(m!)^2}|\operatorname{per}((u_i,v_j))|^2,$$

$$\|u_1*\cdots*u_m\|^2 = \frac{1}{m!}\operatorname{per}((u_i,u_j)),$$

$$\|v_1*\cdots*v_m\|^2 = \frac{1}{m!}\operatorname{per}((v_i,v_j)),$$

which together with (2.11) yield inequality (2.9). Also, equality can hold in (2.11) if and only if the tensors $u_1*\cdots*u_m$ and $v_1*\cdots*v_m$ are linearly dependent. The conditions for equality in (2.9) now follow by Theorem 2.3.

(b) Set $u_i = A_{(i)}$, $i=1,\ldots,m$, and $v_j = \overline{B}^{(j)}$, $j=1,\ldots,m$, and the result follows by Part (a). ■

Corollary 1. *If* A *is an* n*-square matrix, then*

$$|\operatorname{per}(A)|^2 \leqslant \operatorname{per}(AA^*). \tag{2.12}$$

Equality holds in (2.12) *if and only if either* A *has a zero row, or* A *is a generalized permutation matrix.*

Proof. Set $B = I_n$ in (2.10). ■

Corollary 2. *If* U *is a unitary matrix, then*

$$|\operatorname{per}(U)| \leqslant 1. \tag{2.13}$$

The inequality in Theorem 2.4 provides a simple proof for a classical theorem of Schur [34] (see Theorem 4.2 in Chapter 1).

ISBN 0-201-13505-1

THEOREM 2.5 (Schur). *If A is a positive semi-definite hermitian matrix, then*

$$\operatorname{per}(A) \geqslant \det(A), \tag{2.14}$$

with equality if and only if A is diagonal or A has a zero row.

Proof. Since A is positive semi-definite, there exists a triangular matrix T such that $A = TT^*$. Thus

$$\begin{aligned}\det(A) &= \det(TT^*)\\ &= \det(T)\det(T^*).\end{aligned}$$

But since T is triangular, $\det(T) = \operatorname{per}(T)$ and $\det(T^*) = \operatorname{per}(T^*)$. Hence

$$\det(A) = \operatorname{per}(T)\operatorname{per}(T^*)$$

and, by Theorem 2.4,

$$\begin{aligned}\operatorname{per}(T) &\leqslant |\operatorname{per}(TI_n)|\\ &\leqslant \sqrt{\operatorname{per}(TT^*)}\\ &= \sqrt{\operatorname{per}(A)}\ ,\\ \operatorname{per}(T^*) &\leqslant |\operatorname{per}(I_n T^*)|\\ &\leqslant \sqrt{\operatorname{per}(TT^*)}\\ &= \sqrt{\operatorname{per}(A)}\ .\end{aligned}$$

The result follows. ■

Problems

1. Let

$$A = \begin{bmatrix} 2 & 3 & 1 \\ 1 & 2 & 2 \end{bmatrix} \quad \text{and} \quad B = \begin{bmatrix} 3 & 1 \\ 1 & 2 \\ 1 & 3 \end{bmatrix}.$$

   Use the Binet–Cauchy theorem to evaluate the permanent of AB.
2. Prove Theorem 1.2.
3. Let A be an $m \times n$ matrix, D an $m \times m$ diagonal matrix, and G an $n \times n$ diagonal matrix. Show that $\operatorname{Per}(DAG) \neq \operatorname{Per}(D)\operatorname{Per}(A)\operatorname{Per}(G)$, in general.

ISBN 0-201-13505-1

4. Let $A=(a_{ij})\in M_n$ be a (0,1)-matrix—that is, a matrix each of whose entries is either 0 or 1. Let $B=(b_{ij})$ be the permanental adjugate of A—that is, $b_{ij}=\mathrm{per}(A(j|i))$. Show that

$$(\mathrm{per}(A))^2\leqslant a\,\mathrm{tr}(BB^*),$$

where $a=\sum_{i,j}a_{ij}/n^2$.

5. Prove that, if $A\in M_n$, then

$$\mathrm{per}(\lambda I_n-A)=\lambda^n+\sum_{k=1}^{n}c_k\lambda^{n-k},$$

where $c_k=(-1)^k\sum_{\omega\in Q_{k,n}}\mathrm{per}(A[\omega])$. (Note: $\mathrm{per}(\lambda I_n-A)$ is called the *permanental characteristic polynomial* of A.)

6. Given $C=A+iB$, where A and B are 3×3 real matrices, use Theorem 1.4 to evaluate the real and the imaginary parts of $\mathrm{per}(C)$.

7. Let $A\in M_n(\mathbf{C})$ and $A=HU$, where U is unitary and H is positive semi-definite hermitian. Show that

$$|\mathrm{per}(A)|\leqslant(\mathrm{per}(H^2))^{\frac{1}{2}}.$$

8. If A is a complex matrix with row sums $r_1,\ldots,r_n$, show that

$$\mathrm{per}(AA^*)\geqslant\frac{n!}{n^n}\prod_{i=1}^{n}|r_i|^2.$$

9. Prove Corollary 2 to Theorem 2.4.

ISBN 0-201-13505-1

CHAPTER 3

(0, 1)-Matrices

3.1 Incidence Matrices

Matrices all of whose entries are either 0 or 1—that is, (0, 1)-matrices—play an important part in linear algebra, combinatorics, and graph theory. In some of these applications it is at times preferable to consider 1 as the "all" element in a Boolean algebra, or the identity element in a field of two elements. In what follows, however, the symbol 1 will represent the positive integer 1, since we shall be mainly concerned with enumerations of systems of distinct representatives and with related problems in the theory of permanents.

Many problems in the theory of nonnegative matrices depend only on the distribution of zero entries. In such cases the relevant property of each entry is whether it is zero or nonzero, and the problem can be often simplified by substituting for the given matrix the (0, 1)-matrix with exactly the same zero pattern.

DEFINITION 1.1. Two $m \times n$ matrices $A=(a_{ij})$ and $B=(b_{ij})$ are said to have the *same zero pattern* if $a_{ij}=0$ implies $b_{ij}=0$, and vice versa.

Suppose that A, B, C, and D are nonnegative n-square matrices, and that A has the same zero pattern as B, and C has the same zero pattern as D. Then clearly $A+C$ has the same zero pattern as $B+D$, and AC has the same zero pattern as BC.

Many problems of a combinatorial nature can be translated into problems involving configurations of subsets of a finite set, which in turn can be conveniently represented by (0, 1)-matrices.

DEFINITION 1.2. Let $X_1, X_2, \ldots, X_m$ be subsets, not necessarily distinct, of an n-set $S=\{x_1, x_2, \ldots, x_n\}$. Let $A=(a_{ij})$ be the (0, 1)-matrix whose entries

ISBN 0-201-13505-1

ENCYCLOPEDIA OF MATHEMATICS and Its Applications, Gian-Carlo Rota (ed.). Vol. 6: Henryk Minc, Permanents

are defined as follows:

$$a_{ij} = \begin{cases} 1 & \text{if } x_j \in X_i, \\ 0 & \text{if } x_j \notin X_i. \end{cases}$$

The matrix A is called the *incidence matrix* for the subsets $X_1, X_2, \ldots, X_m$ of the n-set S.

The concept of incidence matrices can be specialized to provide representations for various relations, functions, graphs, set intersections, etc. (See Problems 1–9, and the Dance Problem below.)

DEFINITION 1.3. Again, let $X_1, X_2, \ldots, X_m$ be subsets of an n-set S. A sequence $(s_1, s_2, \ldots, s_m)$ of m distinct elements of S is said to form a *system of distinct representatives*, abbreviated SDR, if $s_i \in X_i$, $i = 1, \ldots, m$.

A configuration of subsets may or may not have an SDR. It is of considerable interest in combinatorics to determine whether a given configuration has an SDR and, if so, how many.

For example, in the well-known Dance Problem we are asked the following question. If at a school dance there are n boys and n girls, and if each boy has previously met exactly k girls and each girl has previously met exactly k boys, is it possible to pair off boys and girls into dance partners who had been previously introduced? In how many ways can it be done?

Let X_i be the subset of girls who had been introduced to the ith boy, $i = 1, \ldots, n$. We are asked if the configuration has an SDR and, if so, how many. In terms of incidence matrices the configuration may be represented by an n-square (0, 1)-matrix with exactly k 1's in each row and each column where the (i,j) entry is 1 if the ith boy has been introduced to the jth girl. The question now is whether it is possible to choose n positions in the matrix, no two in the same row or the same column, with entries equal to 1. We shall see that this question can be answered in the affirmative. The second question asks in how many ways can this be done. The answer to this question is not known. There is not even a reasonable conjecture for a lower bound for this number.

Let $A = (a_{ij})$ be the incidence matrix for subsets $X_1, X_2, \ldots, X_m$ of an n-set $\{x_1, x_2, \ldots, x_n\}$. If the configuration has an SDR, then clearly $m \leqslant n$, and there exists a one-to-one function σ: $\{1, \ldots, m\} \to \{1, \ldots, n\}$ such that

$$x_{\sigma(i)} \in X_i,$$

$i = 1, \ldots, m$. It follows from the definition of incidence matrices that

$$a_{i\sigma(i)} = 1,$$

ISBN 0-201-13505-1

$i=1,\ldots,m$. Hence the configuration has an SDR if and only if there exists a one-to-one function σ such that $a_{1\sigma(1)}=a_{2\sigma(2)}=\cdots=a_{m\sigma(m)}=1$; i.e.,

$$\prod_{i=1}^{m} a_{i\sigma(i)}=1. \tag{1.1}$$

The number of SDR's is equal to the number of one-to-one functions σ for which (1.1) holds. It is therefore equal to

$$\operatorname{Per}(A)=\sum_{\sigma}\prod_{i=1}^{m} a_{i\sigma(i)}.$$

Example 1.1. Find the number of SDR's in the following configuration of subsets of $\{x_1,x_2,x_3,x_4\}$: $X_1=\{x_1,x_3\}$, $X_2=\{x_1,x_2\}$, $X_3=\{x_2,x_3,x_4\}$. The incidence matrix of the configuration is

$$A=\begin{bmatrix}1&0&1&0\\1&1&0&0\\0&1&1&1\end{bmatrix}.$$

The number of SDR's is $\operatorname{Per}(A)=5$. In fact, the SDR's are: (x_1,x_2,x_3), (x_1,x_2,x_4), (x_3,x_1,x_2), (x_3,x_1,x_4), and (x_3,x_2,x_4).

3.2 Theorems of Frobenius and König

The most fundamental result in the combinatorial matrix theory and the theory of the permanents of nonnegative matrices is the so-called Frobenius–König theorem. It was first obtained by Frobenius, then re-proved by König by elementary graph-theoretical methods, and again re-proved by an elementary method by Frobenius. A controversy developed between Frobenius and König about this theorem and its relation to other results of König (see [48]). We shall not attempt to adjudicate in this matter and shall refer to the result in question (see below) as the Frobenius–König theorem, by which name it is generally known.

Theorem 2.1 (Frobenius–König). *Let A be an n-square matrix. A necessary and sufficient condition for every diagonal of A to contain a zero entry is that A contain an $s\times t$ zero submatrix such that $s+t=n+1$.*

The proof of Theorem 2.1 can be found in many books (e.g., [117]). Here we prove the following result, which clearly is equivalent to Theorem 2.1.

Theorem 2.2. *Let A be an n-square nonnegative matrix. Then* $\operatorname{per}(A)=0$ *if and only if A contains an $s\times t$ zero submatrix such that $s+t=n+1$.*

ISBN 0-201-13505-1

Proof. We prove the necessity by induction on n. Let $\operatorname{per}(A)=0$. If $n=1$, then $A=[0]$. Assume that the condition is necessary for all square (0,1)-matrices of order less than n. If every entry of A is zero, the proof of necessity is obviously finished. Suppose that $a_{hk}>0$. Then

$$0=\operatorname{per}(A)=\sum_{j=1}^{n} a_{hj}\operatorname{per}(A(h|j)),$$

and, since all the products $a_{hj}\operatorname{per}(A(h|j))$ are nonnegative and a_{hk} is positive, we must have

$$\operatorname{per}(A(h|k))=0.$$

Hence, by the induction hypothesis, we can find an $s_1\times t_1$ zero submatrix of $A(h|k)$ such that $s_1+t_1=(n-1)+1$. Permute the rows and columns of A so that the zero submatrix appears in the top right corner.

$$B=PAQ=\begin{array}{c} \begin{array}{cc} \overbrace{\qquad}^{n-t_1} & \overbrace{\qquad}^{t_1} \end{array} \\ \left[\begin{array}{c|c} X & 0 \\ \hline Z & Y \end{array}\right]\begin{array}{l} \}s_1 \\ \}n-s_1 \end{array} \end{array}.$$

(This permutation of rows and columns is performed solely for notational convenience and is not an essential part of the proof.) Now, neither $n-s_1$ nor $n-t_1$ can be zero, because s_1 and t_1 are positive integers and $s_1+t_1=n$. Since $n-s_1=t_1$ and $n-t_1=s_1$, it follows that X is s_1-square and Y is t_1-square. Thus,

$$0=\operatorname{per}(A)=\operatorname{per}(X)\operatorname{per}(Y),$$

and either $\operatorname{per}(X)$ or $\operatorname{per}(Y)$ must be zero. We can assume without loss of generality that $\operatorname{per}(X)=0$. Then by the induction hypothesis, X contains a $u\times v$ zero submatrix such that $u+v=s_1+1$. Suppose that $X[\alpha_1,\ldots,\alpha_u|\,\beta_1,\ldots,\beta_v]=0$; that is, $B[\alpha_1,\ldots,\alpha_u|\,\beta_1,\ldots,\beta_v]=0$. Consider the submatrix

$$C=B[\alpha_1,\ldots,\alpha_u|\,\beta_1,\ldots,\beta_v,s_1+1,s_1+2,\ldots,n].$$

It is a zero submatrix of A with u rows and $v+(n-s_1)$ columns. Moreover,

$$\begin{aligned} u+v+(n-s_1)&=(u+v)+n-s_1\\ &=s_1+1+n-s_1\\ &=n+1. \end{aligned}$$

ISBN 0-201-13505-1

This completes the proof of necessity. To prove the converse, suppose that A contains an $s\times t$ zero submatrix with $s+t=n+1$. Permute the rows and columns of A so that

$$B=PAQ=\begin{array}{c} \\ s\left\{\right. \\ \\ \end{array}\left[\begin{array}{c|c} \overbrace{0}^{t} & K \\ \hline L & M \end{array}\right]. \tag{2.1}$$

By the Laplace expansion theorem (Theorem 1.2, Chapter 2),

$$\operatorname{per}(A)=\operatorname{per}(B)=\sum_{\omega\in Q_{s,n}}\operatorname{per}(B[1,\ldots,s|\omega])\operatorname{per}(B(1,\ldots,s|\omega)). \tag{2.2}$$

But $B[1,\ldots,s|\omega]$ is an $s\times s$ submatrix of $B[1,\ldots,s|1,\ldots,n]$, and the latter matrix has at most

$$n-t=s+t-1-t=s-1$$

nonzero columns. Hence every submatrix $B[1,\ldots,s|\omega]$ must have at least one zero column, and therefore $\operatorname{per}(B[1,\ldots,s|\omega])=0$ for all $\omega\in Q_{s,n}$. It follows from (2.2) that $\operatorname{per}(A)=0$. ∎

The following theorem is a slightly extended version of Theorem 2.2.

THEOREM 2.3. *Let A be a nonnegative $m\times n$ matrix, $m\leqslant n$. Then* $\operatorname{Per}(A)=0$ *if and only if A contains an $s\times(n-s+1)$ zero submatrix.*

Proof. If $m<n$, let

$$B=\begin{bmatrix} A \\ C \end{bmatrix},$$

where C is an $(n-m)\times n$ matrix, all of whose entries are 1. Then, by Theorem 2.2, $\operatorname{per}(B)=0$ if and only if B (and therefore A) contains a zero $s\times(n-s+1)$ submatrix. Also, $\operatorname{per}(B)=(n-m)!\operatorname{Per}(A)$ is 0 if and only if $\operatorname{Per}(A)=0$. ∎

An important concept in the combinatorial theory of matrices is that of the term rank of a (0,1)-matrix.

DEFINITION 2.1. Let A be an $m\times n$ (0,1)-matrix. The *term rank* of A is the maximal order of a square submatrix of A with a nonzero permanent. Alternatively, the term rank of A is the maximal number of 1's in a diagonal of A.

The second version of the definition can be generalized in an obvious way to matrices over any field.

ISBN 0-201-13505-1

The key theorem on term rank is the König–Egerváry theorem [41, 42, 48]. Recall that a *line* of a matrix designates either its row or its column.

THEOREM 2.4 (König–Egerváry). *Let A be an $m\times n$ (0, 1)-matrix. The term rank of A is equal to the minimal number of lines in A that contain all the 1's in A.*

The proof of this classical theorem is quite straightforward (see, e.g., [87]).

Another version of this result is a generalization of the Frobenius–König theorem due to König [48].

THEOREM 2.5. *Let A be an $m\times n$ (0, 1)-matrix, $m\leqslant n$. The term rank of A is ρ if and only if A contains an $s\times t$ zero submatrix with $s+t=n+m-\rho$ but does not contain any $p\times q$ zero submatrix such that $p+q>n+m-\rho$.*

3.3 Structure of Square (0, 1)-Matrices

The permanent of a (0, 1)-matrix depends on the zero pattern of the matrix. In this section we study structural properties of square (0, 1)-matrices. We shall see that one of the relevant properties of a (0, 1)-matrix is the existence or nonexistence of a doubly stochastic matrix of the same zero pattern. We begin with some basic properties of doubly stochastic matrices.

Recall that a nonnegative matrix is called *doubly stochastic* if all its row sums and column sums are 1. The set of all $n\times n$ doubly stochastic matrices is denoted by Ω_n.

DEFINITION 3.1. An n-square nonnegative matrix is said to be *partly decomposable* if it contains a $k\times(n-k)$ zero submatrix. In other words, a matrix A is partly decomposable if there exist permutation matrices P and Q such that

$$PAQ=\begin{bmatrix} B & C \\ 0 & D \end{bmatrix},$$

where B and D are square. If the matrix contains no $k\times(n-k)$ zero submatrix for $k=1,\ldots,n-1$, it is called *fully indecomposable.*

THEOREM 3.1. *If A is a partly decomposable doubly stochastic matrix, then there exist permutation matrices P and Q such that PAQ is a direct sum of doubly stochastic matrices.*

For, suppose that

$$PAQ=\begin{bmatrix} B & C \\ 0 & D \end{bmatrix},$$

ISBN 0-201-13505-1

where D is k-square. Since PAQ is doubly stochastic, the sum of entries in the first $n-k$ columns of PAQ is $n-k$, and

$$s(B)=n-k,$$

where $s(X)$ denotes the sum of entries in the matrix X. Similarly, considering the entries in the last k rows of PAQ, we can conclude that

$$s(D)=k.$$

But

$$\begin{aligned} n &= s(PAQ) \\ &= s(B)+s(C)+s(D) \\ &= n-k+s(C)+k \\ &= n+s(C), \end{aligned}$$

and therefore

$$s(C)=0.$$

Since C is nonnegative, we must have $C=0$, and thus $PAQ=B\dotplus D$, where B and D are clearly doubly stochastic.

THEOREM 3.2. *The permanent of a doubly stochastic matrix is positive.*

For, if $\operatorname{per}(A)=0$, then, by the Frobenius–König theorem, there exist permutation matrices P and Q such that

$$PAQ=\begin{bmatrix} B & C \\ 0 & D \end{bmatrix},$$

where the zero submatrix in the bottom left corner is $h\times k$, with $h+k=n+1$. But if $A\in\Omega_n$, then

$$\begin{aligned} n &= s(PAQ) \\ &\geqslant s(B)+s(D). \end{aligned}$$

Now, all the nonzero entries in the first k columns are contained in B, and thus $s(B)=k$. Similarly, $s(D)=h$. Hence

$$\begin{aligned} n &\geqslant s(B)+s(D) \\ &= k+h, \end{aligned}$$

which is impossible, since $h+k=n+1$.

ISBN 0-201-13505-1

COROLLARY. *Every doubly stochastic matrix has a positive diagonal.*

Note that the corollary answers in the affirmative the question raised in the Dance Problem (see Section 3.1). For, if $S=(s_{ij})$ is the incidence matrix of the configuration (that is, $s_{ij}=1$ if the ith boy has been introduced to the jth girl, and $s_{ij}=0$ otherwise), then $\frac{1}{k}S$ is doubly stochastic. Hence S has a positive diagonal. In other words, it is possible to pair off boys and girls as required.

We now introduce one of the fundamental results in the theory of doubly stochastic matrices.

THEOREM 3.3 (Birkhoff, 1946). *The set Ω_n of n-square doubly stochastic matrices forms a convex polyhedron with permutation matrices as vertices. In other words, if $A\in\Omega_n$, then*

$$A=\sum_{j=1}^{s}\theta_j P_j,$$

where $P_1,\ldots,P_s$ are permutation matrices and $\theta_1,\ldots,\theta_s$ are nonnegative numbers, $\sum_{j=1}^{s}\theta_j=1$.

Proof. Use induction on $\pi(A)$, the number of positive entries in A. If $\pi(A)=n$, then A is a permutation matrix, and the theorem holds ($s=1$). Assume that $\pi(A)>n$ and that the theorem holds for all matrices in Ω_n with less than $\pi(A)$ positive entries. By the corollary to Theorem 3.2, the matrix A has a positive diagonal $(a_{\sigma(1),1},a_{\sigma(2),2},\ldots,a_{\sigma(n),n})$. Let $a_{\sigma(t),t}=\min_i(a_{\sigma(i),i})=a$, and let $P=(p_{ij})$ be the incidence matrix of the permutation σ (that is, P is the permutation matrix with 1's in positions $(\sigma(i),i)$, $i=1,\ldots,n$). Clearly $0<a<1$, since $a=1$ would imply that A has 1's in the positions $(\sigma(i),i)$ and is therefore a permutation matrix. Also, $A-aP$ is nonnegative because of the minimality of a. We assert that

$$B=(b_{ij})=\frac{1}{1-a}(A-aP) \tag{3.1}$$

is doubly stochastic. Indeed,

$$\begin{aligned}\sum_{j=1}^{n}b_{ij}&=\frac{1}{1-a}\sum_{j=1}^{n}(a_{ij}-ap_{ij})\\&=\frac{1}{1-a}\left[\left(\sum_{j=1}^{n}a_{ij}\right)-a\sum_{j=1}^{n}p_{ij}\right]\\&=\frac{1}{1-a}(1-a)\\&=1,\end{aligned}$$

ISBN 0-201-13505-1

$i=1,\ldots,n$, and we can show similarly that

$$\sum_{i=1}^{n} b_{ij}=1,$$

$j=1,\ldots,n$. Now, $\pi(B)\leqslant\pi(A)-1$, since B has zeros in all positions where A has a zero entry, and in addition $b_{\sigma(t),t}=0$. Hence, by the induction hypothesis,

$$B=\sum_{j=1}^{s-1}\varphi_j P_j,$$

where the P_j are permutation matrices, $\varphi_j\geqslant 0, j=1,\ldots,s-1$, and $\sum_{j=1}^{s-1}\varphi_j=1$. But then, by (3.1),

$$\begin{aligned}A&=(1-a)B+aP\\&=\left(\sum_{j=1}^{s-1}(1-a)\varphi_j P_j\right)+aP\\&=\sum_{j=1}^{s}\theta_j P_j,\end{aligned}$$

where $\theta_j=(1-a)\varphi_j, j=1,\ldots,s-1$, $\theta_s=a$, and $P_s=P$. Obviously, the θ_j are nonnegative. It remains to show that $\sum_{j=1}^{s}\theta_j=1$. We compute

$$\begin{aligned}\sum_{j=1}^{s}\theta_j&=\left(\sum_{j=1}^{s-1}(1-a)\varphi_j\right)+a\\&=(1-a)\sum_{j=1}^{s-1}\varphi_j+a\\&=1-a+a\\&=1.\end{aligned}$$

■

Let Λ_n^k denote the set of n-square (0, 1)-matrices with k 1's in each row and each column. Of course, the set Λ_n^k is not convex. However, if $A\in\Lambda_n^k$, then $\frac{1}{k}A\in\Omega_n$, and thus matrices in Λ_n^k virtually have the properties of doubly stochastic matrices. Matrices in Λ_n^k are therefore usually called *doubly stochastic* (0, 1)-*matrices*. The following result is the analogue of Theorem 3.3, although historically it preceded Birkhoff's theorem.

ISBN 0-201-13505-1

THEOREM 3.4 [48]. *If $A \in \Lambda_n^k$ then*

$$A = \sum_{j=1}^{k} P_j,$$

where the P_j are permutation matrices.

The proof of Theorem 3.4 follows the lines of that of Theorem 3.3, but it is much simpler. We leave it as an exercise (Problem 12).

If A is an incidence matrix of a configuration, then permutations of rows and columns of A correspond to relabeling the subsets and the elements of the configuration, respectively. Also, the permanent of A is invariant under permutations of rows and columns of A. In other words, the combinatorial essentials of the configuration are not changed by pre- and postmultiplying by permutation matrices.

Recall that an n-square nonnegative matrix A is fully indecomposable if there exist no permutation matrices such that

$$PAQ = \begin{bmatrix} B & C \\ 0 & D \end{bmatrix},$$

where the zero block is $s \times (n-s)$. The concept of a fully indecomposable matrix plays in the combinatorial matrix theory a part similar to that of an irreducible matrix in the spectral theory of nonnegative matrices.

THEOREM 3.5. *A nonnegative n-square matrix A, $n \geqslant 2$, is fully indecomposable if and only if*

$$\operatorname{per}(A(i|j)) > 0$$

for all i and j.

Proof. By Theorem 2.1, $\operatorname{per}(A(h|k)) = 0$ for some h and k if and only if the submatrix $A(h|k)$, and thus the matrix A, contains an $s \times t$ zero submatrix with $s + t = (n-1) + 1$. In other words, $\operatorname{per}(A(h|k)) = 0$ for some h and k if and only if the matrix A is partly decomposable. ■

Let E_{ij} denote the $n \times n$ matrix with 1 in the (i,j) position and zeros elsewhere. The following result is an immediate consequence of Theorem 3.5.

THEOREM 3.6. *If A is a fully indecomposable matrix and $c \neq 0$, then for every i and j,*

$$\operatorname{per}(A + cE_{ij}) > \operatorname{per}(A)$$

ISBN 0-201-13505-1

or

$$\operatorname{per}(A+cE_{ij})<\operatorname{per}(A),$$

according as $c>0$ *or* $c<0$.

For, $\operatorname{per}(A+cE_{ij})=\operatorname{per}(A)+c\operatorname{per}(A(i|j))$ and, by Theorem 3.5, $\operatorname{per}(A(i|j))>0$.

A stronger result is possible in the case of a fully indecomposable (0, 1)-matrix.

THEOREM 3.7. *If A is a fully indecomposable* (0, 1)*-matrix, then*

$$\operatorname{per}\left(A+\sum_{t=1}^{m}E_{i_tj_t}\right)\geqslant\operatorname{per}(A)+m.$$

Proof. Since A is a (0, 1)-matrix, Theorem 3.5 implies that

$$\operatorname{per}(A(i|j))\geqslant 1$$

for all i and j. Therefore

$$\begin{aligned}\operatorname{per}(A+E_{i_1j_1})&=\operatorname{per}(A)+\operatorname{per}(A(i_1|j_1))\\&\geqslant\operatorname{per}(A)+1.\end{aligned}$$

Clearly $A+E_{i_1j_1}$ is fully indecomposable. The result now follows by induction on m. ■

THEOREM 3.8. *Let*

$$A=\begin{bmatrix}A_1 & B_1 & 0 & & \cdots & 0\\ 0 & A_2 & B_2 & & & \vdots\\ & & \cdot & \cdot & & \\ \vdots & & & \cdot & \cdot & 0\\ & & & \cdot & \cdot & \\ 0 & 0 & & & A_{r-1} & B_{r-1}\\ B_r & 0 & \cdots & & 0 & A_r\end{bmatrix}\tag{3.2}$$

be a nonnegative $n\times n$ matrix, where A_i is a fully indecomposable $n_i\times n_i$ matrix, $i=1,\ldots,r$, and $B_i\neq 0$, $i=1,\ldots,r$. Then A is fully indecomposable.

Proof [G. Edgar, unpublished]. Suppose that A is partly decomposable—i.e., that $A[\alpha|\beta]=0$ for some $\alpha\in Q_{s,n}$ and $\beta\in Q_{t,n}$, where $s+t=n$. Let s_j of rows α and t_j of columns β intersect the submatrix A_j,

ISBN 0-201-13505-1

$j=1,\ldots,r$. Then $s_1+s_2+\cdots+s_r=s>1$, so that at least one of the s_j must be positive. Similarly, at least one of the t_j is not zero. Now, since each A_j is fully indecomposable and it contains an $s_j\times t_j$ zero submatrix (unless either $s_j=0$ or $t_j=0$), we must have $s_j+t_j\leqslant n_j$ where equality can hold only if $s_j=0$ or $t_j=0$. But

$$\begin{aligned} n &= s+t \\ &= \sum_{j=1}^{r} s_j + \sum_{j=1}^{r} t_j \\ &= \sum_{j=1}^{r} (s_j+t_j) \\ &\leqslant \sum_{j=1}^{r} n_j \\ &= n, \end{aligned}$$

and thus $s_j+t_j=n_j$ for every j. It follows that either $s_j=0$ or $t_j=0$ for $j=1,\ldots,r$. But not all the s_j nor all the t_j can be zero, and therefore there must exist an integer k such that $s_k=n_k$ and $t_{k+1}=n_{k+1}$ (subscripts reduced modulo r). It follows that B_k is a submatrix of a zero submatrix, contradicting our hypotheses. ■

DEFINITION 3.2. A nonnegative matrix is said to have a *doubly stochastic pattern* if it has the same zero pattern as a doubly stochastic matrix.

For example, the matrix

$$\begin{bmatrix} 1 & 1 & 1 \\ 1 & 1 & 0 \\ 1 & 0 & 1 \end{bmatrix}$$

has a doubly stochastic pattern, since it has the same zero pattern as the doubly stochastic matrix

$$\begin{bmatrix} \frac{1}{2} & \frac{1}{4} & \frac{1}{4} \\ \frac{1}{4} & \frac{3}{4} & 0 \\ \frac{1}{4} & 0 & \frac{3}{4} \end{bmatrix}.$$

On the other hand, the matrix

$$\begin{bmatrix} 1 & 1 & 1 \\ 1 & 1 & 0 \\ 0 & 0 & 1 \end{bmatrix} \tag{3.3}$$

ISBN 0-201-13505-1

does not have a doubly stochastic pattern. For, if any doubly stochastic matrix had the same zero pattern as (3.3), its only nonzero entry in the third column would have to be 1. But then its (1,3) entry could not be positive.

THEOREM 3.9. *A fully indecomposable matrix has a doubly stochastic pattern.*

Proof. Let $A=(a_{ij})$ be an $n\times n$ fully indecomposable matrix. Then, by Theorem 3.5, $\operatorname{per}(A(i|j))>0$ for all i,j. Let $S=(s_{ij})$ be the $n\times n$ matrix defined by

$$s_{ij}=a_{ij}\operatorname{per}(A(i|j))/\operatorname{per}(A),$$

$i,j=1,\ldots,n$. Clearly S is nonnegative, and it has the same zero pattern as A. Also for $i=1,\ldots,n$,

$$\begin{aligned}\sum_{j=1}^{n} s_{ij} &= \frac{1}{\operatorname{per}(A)}\sum_{j=1}^{n} a_{ij}\operatorname{per}(A(i|j))\\ &= \frac{1}{\operatorname{per}(A)}\operatorname{per}(A)\\ &=1,\end{aligned}$$

and similarly, for $j=1,\ldots,n$,

$$\begin{aligned}\sum_{i=1}^{n} s_{ij} &= \frac{1}{\operatorname{per}(A)}\sum_{i=1}^{n} a_{ij}\operatorname{per}(A(i|j))\\ &=1.\end{aligned}$$

Hence S is doubly stochastic, and thus A has a doubly stochastic pattern. ■

We saw in Theorem 3.7 that replacing a zero by a 1 in a fully indecomposable (0, 1)-matrix increases its permanent at least by 1. Clearly, if we reverse the transformation and replace a 1 in a fully indecomposable matrix by a zero, the permanent will decrease at least by 1. We shall consider (0, 1)-matrices obtained by carrying out such a "stripping" process to the limit. Such matrices have a relatively simple structure and allow us to obtain estimates for the permanent of indecomposable (0, 1)-matrices.

DEFINITION 3.3. A nonnegative matrix $A=(a_{ij})$ is said to be *nearly decomposable* if it is fully indecomposable and if it has the property that for each positive entry a_{ij} the matrix $A-a_{ij}E_{ij}$ is partly decomposable.

The following fundamental theorem on the structure of nearly decomposable matrices is due to Sinkhorn and Knopp [170].

ISBN 0-201-13505-1

THEOREM 3.10. *If A is a nearly decomposable nonnegative matrix, then there exist permutation matrices P and Q and an integer $r>1$, such that*

$$PAQ=\begin{bmatrix} A_1 & E_1 & 0 & \cdots & 0 & 0 \\ 0 & A_2 & E_2 & \cdots & 0 & 0 \\ 0 & 0 & A_3 & \cdots & 0 & 0 \\ \vdots & \vdots & \vdots & \ddots & & \vdots \\ 0 & 0 & 0 & & A_{r-1} & E_{r-1} \\ E_r & 0 & 0 & \cdots & 0 & A_r \end{bmatrix}, \tag{3.4}$$

where each E_i has exactly one positive entry and each block A_i is nearly decomposable.

Proof. Let A be nearly decomposable. Then by Theorem 3.9 there exists a doubly stochastic matrix $S=(s_{ij})$ with the same zero pattern as A. Let the least positive entry in S be c; let $s_{hk}=c$. Since S is nearly decomposable, $S-s_{hk}E_{hk}$ is partly decomposable; that is, there exist permutation matrices P_1 and Q_1 such that

$$P_1(S-s_{hk}E_{hk})Q_1=\begin{array}{c} \\ s\{ \end{array}\left[\begin{array}{c|c} X & Y \\ \hline \underbrace{0}_{t} & Z \end{array}\right],$$

where the zero block is $s\times t$, and $s+t=n$. In other words,

$$P_1SQ_1=\begin{array}{c} \\ s\{ \end{array}\left[\begin{array}{c|c} X & Y \\ \hline \underbrace{F}_{t} & Z \end{array}\right],$$

where F has exactly one nonzero element, which is equal to c. Now, P_1SQ_1 is doubly stochastic, and therefore

$$\begin{aligned} n=s(P_1SQ_1)&=s(X)+s(F)+s(Z)+s(Y) \\ &=(s-c)+c+(t-c)+s(Y) \\ &=n-c+s(Y), \end{aligned}$$

ISBN 0-201-13505-1

where $s(M)$ denotes the sum of entries in the matrix M. Hence $s(Y)=c$, and, because of the minimality of c, Y must contain exactly one positive entry. If both X and Z are fully indecomposable, we are finished, as P_1SQ_1 and P_1AQ_1 have the same zero pattern. If X or Z is partly decomposable, then by permuting rows and columns, S can be put in the form

$$PSQ=\left[\begin{array}{ccccc|ccccc}
A_1 & B_{12} & B_{13} & \cdots & B_{1p} & & & & & \\
0 & A_2 & B_{23} & \cdots & B_{2p} & & & & & \\
0 & 0 & A_3 & \cdots & B_{3p} & & & Y_2 & & \\
\vdots & \vdots & \vdots & \ddots & \vdots & & & & & \\
0 & 0 & 0 & \cdots & A_p & & & & & \\
\hline
 & & & & & A_{p+1} & B_{p+1,p+2} & B_{p+1,p+3} & \cdots & B_{p+1,r} \\
 & & & & & 0 & A_{p+2} & B_{p+2,p+3} & \cdots & B_{p+2,r} \\
 & & & & & 0 & 0 & A_{p+3} & \cdots & B_{p+3,r} \\
 & & Y_1 & & & \vdots & \vdots & \vdots & \ddots & \vdots \\
 & & & & & 0 & 0 & 0 & \cdots & A_r
\end{array}\right],$$

where P and Q are permutation matrices, the A_i are nearly decomposable $n_i\times n_i$ matrices, and each of Y_1 and Y_2 has exactly one positive entry, which is equal to c. Clearly the positive entry of Y_1 must lie in the bottom left $n_r\times n_1$ block. For, otherwise, either the first n_1 columns would contain an $(n-n_1)\times n_1$ zero submatrix, or the last n_r rows would contain an $n_r\times(n-n_r)$ zero submatrix, and PSQ (and therefore S) would be partly decomposable. For similar reasons the only positive entry of Y_2 must lie in the rows containing A_p and in the columns containing A_{p+1}. Now PSQ is doubly stochastic, and therefore one column sum of A_1 is $1-c$ and the other column sums are 1. The minimality of c implies that exactly one row sum of A_1 is $1-c$ and all the other row sums are 1. Thus, again due to the minimality of c, all but one of the blocks $B_{12},B_{13},\ldots,B_{1p}$ must be zero, and one block must have exactly one positive entry, which is equal to c. It is easy to see that the nonzero block must be B_{12}; otherwise, PSQ would contain an $(n-n_2)\times n_2$ zero submatrix and would be partly decomposable. It then follows by an analogous argument that B_{23} has exactly one positive entry and that the blocks $B_{24},\ldots,B_{2p}$ are all zero. Continuing in this manner we show that PSQ, and therefore PAQ, is in the form (3.4). ■

Hartfiel [178] showed that P and Q can be so chosen that the A_i are actually 1-square, $i=1,\ldots,r-1$.

ISBN 0-201-13505-1

3.4 (0, 1)-Circulants

Let P_n, or simply P, denote the $n\times n$ permutation matrix with 1's in positions (1,2), (2,3),...,$(n-1,n)$, $(n,1)$. The matrix

$$\sum_{i=0}^{n-1} c_i P^i,$$

where the c_i are 0 or 1 and $P^0=I_n$, is called a (0, 1)-*circulant*. Such circulants occur in certain classical combinatorial problems: the "problem of derangements" and the "problème des ménages."

The first of these asks the following question: In how many ways can a dance be arranged for n married couples, so that no husband dances with his own wife? In other words, we ask for the number of derangements of n elements—that is, permutations fixing no element. Clearly, the number is equal to the permanent of $J-I_n$, where J is the matrix all of whose entries are 1. Let $D_n=J-I_n$. Then

$$\operatorname{per}(D_n)=\sum_{j=2}^{n}\operatorname{per}(D_n(1|j)).$$

Now, each $(n-1)$-square matrix $D_n(1|j)$ has exactly $n-2$ zeros, no two of which are in the same row or the same column. Hence

$$\operatorname{per}(D_n(1|j))=\operatorname{per}(D_{n-1})+\operatorname{per}(D_{n-2})$$

for $j=2,\ldots,n$. It follows that

$$\operatorname{per}(D_n)=(n-1)(\operatorname{per}(D_{n-1})+\operatorname{per}(D_{n-2})). \tag{4.1}$$

It can be proved by induction using (4.1) that

$$\operatorname{per}(D_n)=n!\left(1-\frac{1}{1!}+\frac{1}{2!}-\frac{1}{3!}+\cdots+(-1)^n\frac{1}{n!}\right). \tag{4.2}$$

A considerably more difficult problem is the "*problème des ménages*": In how many ways can n couples be placed at a round table so that men and women sit in alternate places and no husband sits on either side of his wife? Let the wives be seated in alternate places. For each such seating the husbands can be arranged in

$$U_n=\operatorname{per}(J-I_n-P)$$
$$=\operatorname{per}\left(\sum_{i=2}^{n-1}P^i\right)$$

ISBN 0-201-13505-1

ways. The numbers U_n are called *ménage numbers*. The following formula is due to Touchard [45]:

$$\operatorname{per}\left(\sum_{i=2}^{n-1} P^i\right)=\sum_{i=0}^{n}(-1)^i \frac{2n}{2n-i}\binom{2n-i}{i}(n-i)!. \tag{4.3}$$

For a proof, see [50].

The only other permanents of n-square (0, 1)-circulants for which formulas are available are the permanents of

$$Q(n,k)=\sum_{j=0}^{k-1} P^j \tag{4.4}$$

for small values of k.

We begin with the permanents of certain (0, 1) Toeplitz matrices, which are related to circulants of the form (4.4). For $n \geqslant k-1 \geqslant 2$, let $F(n,k)$ denote the $n\times n$ matrix whose (i,j) entry is 1 if $j-i\in\{-1,0,1,2,\dots,k-2\}$, and 0 otherwise. For example, the submatrix obtained from $Q(n,3)$ by deleting its first column and its last row is $F(n-1,3)$.

Now, it is easy to see that, if $n\geqslant 2(k-1)$, then the expansion of the permanent of $F(n,k)$ by the elements of the first row yields

$$\operatorname{per}(F(n,k))=\sum_{i=1}^{k-1}\operatorname{per}(F(n-i,k)). \tag{4.5}$$

We are ready for the cases $k=3$ and 4.

LEMMA 1. *For* $n\geqslant 4$,

$$\operatorname{per}(Q(n,3))=\operatorname{per}(F(n-1,3))+2\operatorname{per}(F(n-2,3))+2.$$

Proof. For $n\geqslant 5$, expand the permanent by the first two columns:

$$\begin{aligned}\operatorname{per}(Q(n,3))&=1+\operatorname{per}(F(n-3,3))+2\operatorname{per}(F(n-2,3))\\&\quad+\operatorname{per}(F(n-4,3))+\operatorname{per}(F(n-3,3))+1\\&=\operatorname{per}(F(n-1,3))+2\operatorname{per}(F(n-2,3))+2,\end{aligned}$$

by (4.5). For $n=4$ the formula is checked directly. ■

We can deduce the following formulas [94].

ISBN 0-201-13505-1

THEOREM 4.1. *For $n \geqslant 5$,*

$$\operatorname{per}(Q(n,3)) = \operatorname{per}(Q(n-1,3)) + \operatorname{per}(Q(n-2,3)) - 2 \tag{4.6}$$

$$= \left(\frac{1+\sqrt{5}}{2}\right)^n + \left(\frac{1-\sqrt{5}}{2}\right)^n + 2. \tag{4.7}$$

Proof. For $n \geqslant 6$, we have by the lemma and (4.5),

$$\begin{aligned}\operatorname{per}(Q(n-1,3)) + \operatorname{per}(Q(n-2,3)) - 2 \\ &= \big(\operatorname{per}(F(n-2,3)) + 2\operatorname{per}(F(n-3,3)) + 2\big) \\ &\quad + \big(\operatorname{per}(F(n-3,3)) + 2\operatorname{per}(F(n-4,3)) + 2\big) - 2 \\ &= \operatorname{per}(F(n-1,3)) + 2\operatorname{per}(F(n-2,3)) + 2 \\ &= \operatorname{per}(Q(n,3)).\end{aligned}$$

We compute directly that $\operatorname{per}(Q(3,3)) = 6$, $\operatorname{per}(Q(4,3)) = 9$, and $\operatorname{per}(Q(5,3)) = 13$. Hence (4.6) holds for $n = 5$ as well.

Formula (4.7) is obtained by solving the linear difference equation (4.6) and using computed initial values for $\operatorname{per}(Q(n,3))$. ■

The case $k = 4$ is similar:

LEMMA 2. *For $n \geqslant 6$,*

$$\begin{aligned}\operatorname{per}(Q(n,k)) \\ &= 2\big(\operatorname{per}(F(n-1,4)) + 2\operatorname{per}(F(n-2,4)) + 3\operatorname{per}(F(n-3,4)) + 1\big).\end{aligned}$$

Proof. For $n \geqslant 10$ we expand $\operatorname{per}(Q(n,4))$ by the first three columns. After some simplification we obtain

$$\begin{aligned}\operatorname{per}(Q(n,4)) &= 8\operatorname{per}(F(n-3,4)) + 12\operatorname{per}(F(n-4,4)) + 14\operatorname{per}(F(n-5,4)) \\ &\quad + 8\operatorname{per}(F(n-6,4)) + 2\operatorname{per}(F(n-7,4)) + 2 \\ &= 2\operatorname{per}(F(n-1,4)) + 4\operatorname{per}(F(n-2,4)) + 6\operatorname{per}(F(n-3,4)) + 2.\end{aligned}$$

For $n = 6$, 7, 8, and 9, the formula is checked directly. ■

THEOREM 4.2. *For $n \geqslant 7$,*

$$\begin{aligned}\operatorname{per}(Q(n,4)) &= \operatorname{per}(Q(n-1,4)) + \operatorname{per}(Q(n-2,4)) \\ &\quad + \operatorname{per}(Q(n-3,4)) - 4 &(4.8) \\ &= 2(\alpha_1^n + \alpha_2^n + \alpha_3^n + 1), &(4.9)\end{aligned}$$

ISBN 0-201-13505-1

where

$$\alpha_1=\left(\sqrt[3]{19+3\sqrt{33}}+\sqrt[3]{19-3\sqrt{33}}+1\right)/3=1.839286\ldots,$$

and α_2 *and* α_3 *are the other two roots of* $x^3-x^2-x-1=0$.

Proof. It follows from the preceding lemma and (4.5) that, for $n\geqslant 9$,

$$\begin{aligned}
&\operatorname{per}(Q(n-1,4))+\operatorname{per}(Q(n-2,4))+\operatorname{per}(Q(n-3,4))-4\\
&\quad=2\operatorname{per}(F(n-2,4))+4\operatorname{per}(F(n-3,4))+6\operatorname{per}(F(n-4,4))+2\\
&\qquad+2\operatorname{per}(F(n-3,4))+4\operatorname{per}(F(n-4,4))+6\operatorname{per}(F(n-5,4))+2\\
&\qquad+2\operatorname{per}(F(n-4,4))+4\operatorname{per}(F(n-5,4))+6\operatorname{per}(F(n-6,4))+2-4\\
&\quad=2\operatorname{per}(F(n-2,4))+12\operatorname{per}(F(n-3,4))\\
&\qquad+6\operatorname{per}(F(n-4,4))+4\operatorname{per}(F(n-5,4))+2\\
&\quad=6\operatorname{per}(F(n-2,4))+8\operatorname{per}(F(n-3,4))+2\operatorname{per}(F(n-4,4))+2\\
&\quad=2\operatorname{per}(F(n-1,4))+4\operatorname{per}(F(n-2,4))+6\operatorname{per}(F(n-3,4))+2\\
&\quad=\operatorname{per}(Q(n,4)).
\end{aligned}$$

The cases $n=7$ and 8 are verified directly (see the table below).

Formula (4.9) is obtained by solving the difference equation (4.8) in the usual way. ■

It would seem from Theorems 4.1 and 4.2 that a formula of the form

$$\operatorname{per}(Q(n,k))=\sum_{i=1}^{k-1}\operatorname{per}(Q(n-i,k))+\text{constant} \tag{4.10}$$

holds for any k. Unfortunately, this is not the case. It was shown in [94] that formula (4.10) cannot be true for any $k\geqslant 5$. In fact, Metropolis, Stein, and Stein [160] obtained the following result.

THEOREM 4.3. *If* $n\geqslant 15$, *then*

$$\operatorname{per}(Q(n,5))=24+\sum_{i=1}^{10}a_i\operatorname{per}(Q(n-i,5)), \tag{4.11}$$

where $a_1=a_2=a_9=-a_4=-a_7=2$, $a_3=a_8=0$, $a_5=-8$, $a_6=-6$, $a_{10}=1$; *and if* $n\geqslant 21$, *then*

$$\operatorname{per}(Q(n,6))=96+\sum_{i=1}^{15}b_i\operatorname{per}(Q(n-i,6)), \tag{4.12}$$

ISBN 0-201-13505-1

where $b_1=b_2=b_{13}=b_{14}=2$, $b_3=b_{15}=1$, $b_4=0$, $b_5=b_{10}=-b_{11}=-4$, $b_6=-18$, $b_7=-16$, $b_8=-12$, $b_9=-10$, $b_{12}=3$.

We reproduce from [94] a table of values of $\operatorname{per}(Q(n,k))$ for $k\leqslant 7$ and $n\leqslant 15$.

n	$k=3$	4	5	6	7
3	6				
4	9	24			
5	13	44	120		
6	20	80	265	720	
7	31	144	579	1,854	5,040
8	49	264	1,265	4,738	14,833
9	78	484	2,783	12,072	43,387
10	125	888	6,208	30,818	126,565
11	201	1,632	13,909	79,118	369,321
12	324	3,000	31,337	204,448	1,081,313
13	523	5,516	70,985	528,950	3,182,225
14	845	10,144	161,545	1,370,674	9,411,840
15	1,366	18,656	369,024	3,557,408	27,888,139

More comprehensive tables can be found in [160].

We conclude the section with the following remarkable result due to Tinsley [69]. Let C be the 7-square (0,1)-circulant I_n+P+P^3. A (0,1)-matrix $A=\sum_{t=1}^{s}Q_t$, where the Q_t's are permutation matrices, is said to be *abelian*, if $Q_iQ_j=Q_jQ_i$ for each i and j.

THEOREM 4.4. *Let A be an $n\times n$ abelian* (0,1)*-matrix with s 1's in each row and column, $s\geqslant 3$. Then*

$$\operatorname{per}(A)=|\det(A)|,$$

if and only if $s=3$, $n=7r$, and there exist permutation matrices S_1,S_2 such that S_1AS_2 is the direct sum of C taken r times.

Problems

1. Let $S=\{s_1,\ldots,s_n\}$ and $T=\{t_1,\ldots,t_m\}$, and let R be a relation from S to T. Then the *incidence matrix* for R is the $m\times n$ matrix $A(R)=(a_{ij})$, defined by

$$a_{ij}=\begin{cases}1 & \text{if } (s_j,t_i)\in R,\\ 0 & \text{if } (s_j,t_i)\notin R.\end{cases}$$

Construct the incidence matrix for the relation

$$R=\{(s_1,t_3),(s_2,t_1),(s_2,t_2),(s_3,t_1)\},$$

ISBN 0-201-13505-1

from

$$S=\{s_1,s_2,s_3\} \qquad \text{to} \qquad T=\{t_1,t_2,t_3,t_4\}.$$

2. Let $S=\{1,2,3,4\}$ and $T=\{1,4,6,12\}$, and let $s\in S$ and $t\in T$ be R-related if s divides t. Construct the incidence matrix for R.
3. What is a necessary and sufficient condition that an $n\times n$ (0, 1)-matrix be the incidence matrix of an equivalence relation?
4. Let X, Y, and Z be finite sets, and let $f: Y\to Z$ and $g: X\to Y$ be functions. Prove that the incidence matrix of fg is the product of the incidence matrices for f and g:

$$A(fg)=A(f)A(g).$$

[The composite function fg is defined by $fg(x)=f(g(x))$ for all $x\in X$.]
5. Let $S=\{x_1,x_2,x_3,x_4,x_5\}$, and let $X_1=\{x_1,x_2\}$, $X_2=\{x_1,x_4,x_5\}$, and $X_3=\{x_2,x_3,x_4\}$ be subsets of S. Write out all the SDR's of the configuration.
6. Construct the incidence matrix for the configuration in Problem 5, and compute its permanent.
7. Interpret the concept of full indecomposability in the context of incidence matrices for a configuration of subsets $S_1,\ldots,S_n$ of an n-set S.
8. Call an $m\times n$ nonnegative matrix A, $m\leqslant n$, *fully indecomposable* if $\mathrm{Per}(A(i|j))>0$ for $i=1,\ldots,m$, $j=1,\ldots,n$. Interpret this definition in the context of incidence matrices for configurations of subsets. Extend the result in Theorem 3.3 to $m\times n$ nonnegative matrices.
9. Extend the definition of near decomposability to $m\times n$ nonnegative matrices, $m\leqslant n$. Generalize Theorem 3.10 to the class of such matrices.
10. Call an $n\times n$ nonnegative matrix A *k-indecomposable* if $\mathrm{per}(A(\alpha|\beta))>0$ for all α and β in $Q_{k,n}$. Show that if A is fully indecomposable (that is, 1-indecomposable), then AA^{T} is 2-indecomposable.
11. Show that, if A is fully indecomposable, then AA^{T} and $A^{\mathrm{T}}A$ are fully indecomposable. Is the converse true?
12. Prove Theorem 3.4 in detail.
13. Show that a nonnegative matrix is doubly stochastic if and only if both A and A^{T} have the eigenvector $(1,1,\ldots,1)$ for the eigenvalue 1.
14. Use the result in Theorem 3.1 to prove Theorem 3.2.
15. Let $A=(a_{ij})$ be a positive semi-definite doubly stochastic matrix. Show that all the row sums and column sums of $A^{1/2}$ must be 1. Is $A^{1/2}$ always doubly stochastic?
16. Let $V_n=(v_{ij})$ be the n-square (0, 1) Toeplitz matrix with $v_{ij}=1$ if $|i-j|\leqslant 2$, and $v_{ij}=0$ otherwise. Prove that, if $n>5$, then

$$\mathrm{per}(V_n)=2\mathrm{per}(V_{n-1})+2\mathrm{per}(V_{n-3})-\mathrm{per}(V_{n-5}).$$

(See [94, Theorem 5].)

ISBN 0-201-13505-1

CHAPTER 4

Lower Bounds for Permanents

4.1 Marshall Hall's Theorem

Inequalities for determinants, permanents, and other scalar functions of real matrices are of considerable interest in linear algebra. Bounds for permanents in terms of some easily computable functions, such as row sums, are particularly important, since the actual evaluation of permanents presents considerable difficulties.

In this chapter we discuss lower bounds mostly for permanents of nonnegative matrices. First we deal with the permanents of (0, 1)-matrices. The permanent of a (0, 1)-matrix is equal to the number of SDR's in the corresponding configuration (see Section 3.1). Bounds for such permanents therefore have combinatorial significance.

The Frobenius–König theorem (Section 3.2) provides a lower bound of a sort for the permanents of (0, 1)-matrices: If an $m \times n$ (0, 1)-matrix A contains no $s \times (n-s+1)$ zero submatrix, then $\operatorname{Per}(A) \geqslant 1$. The first substantial improvement of this bound was obtained by Marshall Hall [52].

We start with a preliminary result (see [251]).

THEOREM 1.1. *If A is an $m \times n$ (0, 1)-matrix, $m \leqslant n$, and if each row sum of A is greater than or equal to m, then*

$$\operatorname{Per}(A) > 0.$$

Proof. If every row sum of A is greater than or equal to m—that is, every row has at least m positive entries—then the number of zeros in any row cannot exceed $n-m$. Thus, if A contains an $s \times t$ zero submatrix, then $t \leqslant n-m$. But the total number of available rows is m, and therefore $s \leqslant m$.

ENCYCLOPEDIA OF MATHEMATICS and Its Applications, Gian-Carlo Rota (ed.). Vol. 6: Henryk Minc, Permanents

ISBN 0-201-13505-1

Thus,

$$\begin{aligned} s+t &\leqslant m+(n-m) \\ &= n. \end{aligned}$$

Hence, by the Frobenius–König theorem, Per(A) cannot vanish. ■

The following theorem was obtained by Marshall Hall [52]. The method of proof and the extension of the result to the case $t>m$ is due to Mann and Ryser [55].

THEOREM 1.2. *Let A be an $m\times n$ (0,1)-matrix, $m\leqslant n$, with at least t 1's in each row. If $t\geqslant m$, then*

$$\operatorname{Per}(A)\geqslant t!/(t-m)!. \tag{1.1}$$

If $t<m$ and $\operatorname{Per}(A)>0$, *then*

$$\operatorname{Per}(A)\geqslant t!. \tag{1.2}$$

Proof. We prove the theorem by induction on m. By virtue of Theorem 1.1 we can assume that $\operatorname{Per}(A)>0$ for all values of t, $0<t\leqslant n$. If $m=1$, then $t\geqslant m$, and $\operatorname{Per}(A)=t=t!/(t-1)!$. Assume that $m>1$ and that the theorem holds for all matrices with fewer than m rows. The permanent of A is positive, and therefore, by Theorem 2.3, Chapter 3, every submatrix $A[\omega_1,\ldots,\omega_k|1,\ldots,n]$, $\omega\in Q_{k,m}$, $k=1,\ldots,m$, contains at least k nonzero columns. Consider two alternatives. Suppose first that, for some h, $1\leqslant h\leqslant m-1$, and for some sequence $\omega\in Q_{h,m}$, the submatrix $A[\omega_1,\ldots,\omega_h|1,\ldots,n]$ has exactly h nonzero columns. In other words, we assume that

$$PAQ=\begin{array}{c} \\ h\left\{\right. \\ \\ \end{array}\!\!\left[\begin{array}{c|c} \overbrace{\quad B\quad}^{h} & 0 \\ \hline C & D \end{array}\right] \tag{1.3}$$

for some permutation matrices P and Q. In each of the first h rows of PAQ, the t positive entries must be contained in B, and therefore each row of B has at least t 1's and $t\leqslant h\leqslant m-1$. Also,

$$\operatorname{Per}(A)=\operatorname{Per}(PAQ)=\operatorname{Per}(B)\operatorname{Per}(D)>0.$$

Hence $\operatorname{Per}(B)>0$ and $\operatorname{Per}(D)>0$. But the $h\times h$ matrix B satisfies the hypotheses of the theorem. Thus, by induction, $\operatorname{Per}(B)\geqslant t!$. Hence

$$\begin{aligned} \operatorname{Per}(A) &= \operatorname{Per}(B)\operatorname{Per}(D) \\ &\geqslant t!\operatorname{Per}(D) \\ &\geqslant t!. \end{aligned}$$

ISBN 0-201-13505-1

If A cannot be put in the form (1.3), then every $k\times n$ submatrix of A, $1\leqslant k\leqslant m-1$, contains at least $k+1$ nonzero columns. Expand the permanent of A by the first row:

$$\operatorname{Per}(A)=\sum_{j=1}^{n} a_{1j}\operatorname{Per}(A(1|j)). \tag{1.4}$$

Now, every $k\times n$ submatrix of A has at least $k+1$ nonzero columns. Therefore, every $k\times(n-1)$ submatrix of A has at least k nonzero columns, and, by Theorem 2.3, Chapter 3, $\operatorname{Per}(A(1|j))>0$, $j=1,\ldots,n$. Also, every row of $A(1|j)$ contains at least $t-1$ 1's. Hence by the induction hypothesis,

$$\operatorname{per}(A(1|j))\geqslant\begin{cases}(t-1)!, & \text{if } t-1\leqslant m-1,\\ (t-1)!/(t-m)!, & \text{if } t-1\geqslant m-1.\end{cases} \tag{1.5}$$

But $t-1\leqslant m-1$ if $t\leqslant m$, and $t-1\geqslant m-1$ if $t\geqslant m$. Thus, if $t\leqslant m$, then from (1.4) and (1.5),

$$\begin{aligned}\operatorname{per}(A)&\geqslant\sum_{j=1}^{n} a_{1j}(t-1)!\\ &=(t-1)!\sum_{j=1}^{n} a_{1j}\\ &\geqslant t!,\end{aligned}$$

since $\sum_{j=1}^{n} a_{1j}\geqslant t$. Similarly, if $t\geqslant m$, then

$$\begin{aligned}\operatorname{per}(A)&\geqslant\sum_{j=1}^{n} a_{1j}\frac{(t-1)!}{(t-m)!}\\ &=\frac{(t-1)!}{(t-m)!}\sum_{j=1}^{n} a_{1j}\\ &=\frac{t!}{(t-m)!}.\end{aligned}$$

■

4.2 (0, 1)-Matrices

We consider now the permanents of square (0, 1)-matrices. Actually, any result on the permanents of square (0, 1)-matrices can be extended to the case of rectangular matrices by bordering the matrices with rows of 1's. In the preceding section we have shown that the permanent of a square (0, 1)-matrix with at least t 1's in each row is greater than or equal to $t!$,

ISBN 0-201-13505-1

provided that the permanent is nonzero. Without such additional hypotheses, lower bounds for (0, 1)-matrices in terms of row sums or other simple functions of the entries cannot be very sharp, since the value of the permanent of a (0, 1)-matrix depends critically on its zero pattern and not merely on its row sums or on the number of 1's in the matrix. In fact, the permanent of an n-square (0, 1)-matrix may vanish even if as many as n^2-n of its entries are equal to 1. In Section 4.3 we shall obtain bounds for permanents of fully indecomposable (0, 1)-matrices. For general (0, 1)-matrices, Jurkat and Ryser [112] obtained the following lower bound. The proof and the discussion of the case of equality are due to Minc [131]. We require the following definitions.

If $\alpha=(a_1,a_2,\ldots,a_n)$ is a real n-tuple, let $\alpha^*=(a_1^*,a_2^*,\ldots,a_n^*)$ denote the n-tuple α rearranged in a nonincreasing order $a_1^*\geqslant a_2^*\geqslant\cdots\geqslant a_n^*$, and let $\alpha'=(a_1',a_2',\ldots,a_n')$ denote the n-tuple α rearranged in a nondecreasing order $a_1'\leqslant a_2'\leqslant\cdots\leqslant a_n'$. If A is an n-square (0, 1)-matrix with row sums $r_1,\ldots,r_n$, denote by $\bar{A}$ the (0, 1)-matrix whose ith row has the first r_i^* entries equal to 1 and the remaining entries $0, i=1,\ldots,n$. The matrix $\bar{A}$ is called the *maximal matrix* corresponding to A [87]. For example, if

$$A=\begin{bmatrix}0&1&0&0&1\\1&1&0&1&0\\0&0&1&1&1\\1&0&0&1&0\\1&1&0&1&1\end{bmatrix},$$

then

$$\bar{A}=\begin{bmatrix}1&1&1&1&0\\1&1&1&0&0\\1&1&1&0&0\\1&1&0&0&0\\1&1&0&0&0\end{bmatrix}.$$

THEOREM 2.1. *If A is an n-square* (0, 1)-*matrix with row sums $r_1,\ldots,r_n$, then*

$$\operatorname{per}(A)\geqslant\prod_{i=1}^{n}\{r_i+i-n\},\tag{2.1}$$

where $\{r_i+1-n\}=\max(0,r_i+1-n)$. *If* $\operatorname{per}(A)\neq 0$, *then equality holds in* (2.1) *if and only if there exists a permutation matrix P such that $AP=\bar{A}$.*

Proof. If $r_n=0$, there is nothing to prove. Otherwise we can assume without loss of generality that $a_{nj}=1, j=1,\ldots,r_n$. We use induction on n.

ISBN 0-201-13505-1

Expanding the permanent of A by the last row, we have:

$$\begin{aligned}\operatorname{per}(A) &= \sum_{j=1}^{r_n} \operatorname{per}(A(n|j)) \\ &\geqslant \sum_{j=1}^{r_n} \prod_{i=1}^{n-1} \{(r_i - a_{ij}) + i - (n-1)\} \qquad (2.2) \\ &\geqslant \sum_{j=1}^{r_n} \prod_{i=1}^{n-1} \{r_i + i - n\}, \qquad (2.3)\end{aligned}$$

since $1 - a_{ij} \geqslant 0$. Hence

$$\begin{aligned}\operatorname{per}(A) &\geqslant r_n \prod_{i=1}^{n-1} \{r_i + i - n\} \\ &= \prod_{i=1}^{n} \{r_i + i - n\}.\end{aligned}$$

Now, consider the case of equality. Assume that

$$\operatorname{per}(A) = \prod_{i=1}^{n} (r_i + i - n) > 0, \qquad (2.4)$$

and that $a_{nj} = 1, j = 1, \ldots, r_n$, and $a_{nj} = 0, j = r_n + 1, \ldots, n$. Equality (2.4) implies equality in (2.3) and thus that $a_{ij} = 1, i = 1, \ldots, n, j = 1, \ldots, r_n$. In other words, if (2.4) holds, then all the entries in the first r_n columns of A must be 1. It follows that $A(n|1) = \cdots = A(n|r_n)$. Therefore

$$\begin{aligned}0 < \operatorname{per}(A) &= \sum_{j=1}^{r_n} \operatorname{per}(A(n|j)) \\ &= r_n \operatorname{per}(A(n|1)),\end{aligned}$$

and thus $\operatorname{per}(A(n|1)) > 0$. Also by (2.4),

$$\operatorname{per}(A) = r_n \prod_{i=1}^{n-1} (r_i + i - n).$$

Hence,

$$\begin{aligned}\operatorname{per}(A(n|1)) &= \prod_{i=1}^{n-1} (r_i + i - n) \\ &= \prod_{i=1}^{n-1} ((r_i - 1) + i - (n-1)),\end{aligned}$$

ISBN 0-201-13505-1

and, by the induction hypothesis, the matrix $A(n|1)$ is a maximal matrix, possibly with its last $n-r_n$ columns permuted. But all the entries in the first column of A are 1. Also, the first r_n entries in the last row of A are 1, and the other entries are 0. Hence $AP=\bar{A}$ for a suitable permutation matrix P.

We prove the converse by showing that, if $A=\bar{A}$, then

$$\operatorname{per}(A)=\prod_{i=1}^{n}\{r_i+i-n\}. \tag{2.5}$$

(The hypothesis $\operatorname{per}(A)>0$ is not required here.) If $r_n=0$, then both sides of (2.5) are 0. Assume, therefore, that $r_n>0$. Since $A=\bar{A}$ implies that $A(n|1)=\cdots=A(n|r_n)$, we have

$$\begin{aligned}\operatorname{per}(A)&=\sum_{j=1}^{r_n}\operatorname{per}(A(n|j))\\&=r_n\operatorname{per}(A(n|1))\\&=r_n\prod_{i=1}^{n-1}\{(r_i-1)+i-(n-1)\},\end{aligned}$$

by the induction hypothesis, and (2.5) follows. ■

It is easy to see that Theorem 2.1 in general does not give a good lower bound. In fact, if there is any zero in the first row, or if at least i entries in the ith row of the matrix are zero, then (2.1) becomes a completely trivial statement that $\operatorname{per}(A)\geqslant 0$. Nevertheless, the bound in Theorem 2.1 can be somewhat enhanced by the following observation. Since the value of the permanent is invariant under permutation of rows, the bound in (2.1) can be improved by a judicious permutation of rows. In fact, the best bound is obtained if the rows are permuted so that the row sums are in nonincreasing order. To justify this assertion we quote the following rearrangement theorem. (See [202].)

THEOREM 2.2. *If $a=(a_1,\ldots,a_n)$ and $b=(b_1,\ldots,b_n)$ are real n-tuples, and if $a_i+b_i\geqslant 0$, $i=1,\ldots,n$, then*

$$\prod_{i=1}^{n}(a_i'+b_i')\leqslant\prod_{i=1}^{n}(a_i+b_i)\leqslant\prod_{i=1}^{n}(a_i^*+b_i'). \tag{2.6}$$

Moreover, $\prod_{i=1}^{n}(a_i'+b_i')=\prod_{i=1}^{n}(a_i+b_i)\neq 0$ if and only if $a'+b'$ is a rearrangement of $a+b$, and $\prod_{i=1}^{n}(a_i+b_i)=\prod_{i=1}^{n}(a_i^*+b_i')\neq 0$ if and only if $a+b$ is a rearrangement of a^*+b'.

We use this result to prove the following theorem.

ISBN 0-201-13505-1

THEOREM 2.3. *If* $r_1, \ldots, r_n$ *are positive integers not exceeding* n, *then*

$$\prod_{i=1}^{n} \{r_i^* + i - n\} \geqslant \prod_{i=1}^{n} \{r_i + i - n\}. \tag{2.7}$$

Equality can hold in (2.7) *if and only if either the left-hand side of* (2.7) *is* 0, *or* $r_i^* = r_i, i = 1, \ldots, n$.

Proof. If the left-hand side of (2.7) is positive, the result, together with the conditions for equality, holds by virtue of (2.6). It remains to show that, if any of the $\{r_i^* + i - n\}$ are zero, then the right-hand side of (2.7) must vanish. This will also establish the necessity of the conditions for equality; its sufficiency is obvious. Suppose then that $\{r^* + h - n\} = 0$—that is, that $r_h^* + h - n \leqslant 0$, for some h, $1 \leqslant h \leqslant n$. Then we must also have $r_k + k - n \leqslant 0$ for some k, $1 \leqslant k \leqslant h$. For, if $r_h + h - n > 0$, then $r_h > r_h^*$, and therefore there exists an integer k, $1 \leqslant k < h$, such that $r_k \leqslant r_h^*$. It follows that $r_k + h - n \leqslant 0$, and thus, *a fortiori*, that $r_k + k - n < 0$. Hence, if $\{r_h^* + h - n\} = 0$, then both sides of (2.7) vanish. ■

Theorem 2.1 implies that [202]

$$\operatorname{per}(A) \geqslant \prod_{i=1}^{n} \{r_i^* + i - n\}. \tag{2.8}$$

The bound in (2.8) is, by Theorem 2.3, sharper than that in (2.1).

4.3 Fully Indecomposable (0, 1)-Matrices

We saw in Section 3.3 (Theorem 3.5) that the permanents of all $(n-1)$-square submatrices of a fully indecomposable $n \times n$ matrix are positive. This implies the following inequality [236].

THEOREM 3.1. *If* A *is a fully indecomposable n-square* (0, 1)*-matrix with row sums* $r_1, \ldots, r_n$, *then*

$$\operatorname{per}(A) \geqslant \max_i r_i. \tag{3.1}$$

Equality can hold in (3.1) *if and only if at least* $n-1$ *of the row sums equal* 2.

There is, of course, an exact analogue of the theorem involving column sums instead of row sums. Note that the condition for equality does not determine the zero pattern of the matrix (not even modulo permutations of rows and columns) or, in particular, its column sums.

Proof. By Theorem 3.5, Chapter 3, $\operatorname{per}(A(i|j)) \geqslant 1$ for all i and j, since the least nonzero value the permanent of a (0, 1)-matrix may have is 1.

ISBN 0-201-13505-1

Hence

$$\begin{aligned}\operatorname{per}(A) &= \sum_{j=1}^{n} a_{ij}\operatorname{per}(A(i|j))\\ &\geqslant \sum_{j=1}^{n} a_{ij}\\ &= r_i\end{aligned}$$

for all i. The inequality (3.1) follows. We postpone the proof of the case of equality until we establish the following result [162], which gives a considerably sharper bound than the one in (3.1).

THEOREM 3.2. *If* $A=(a_{ij})$ *is a fully indecomposable* n*-square* $(0,1)$*-matrix, then*

$$\operatorname{per}(A) \geqslant s(A)-2n+2, \tag{3.2}$$

where $s(A)$ *denotes the sum of all entries in* A.

Proof. First suppose that A is nearly decomposable. Then, by Theorem 3.10, Chapter 3, there exist permutation matrices P and Q such that

$$PAQ=\begin{bmatrix} A_1 & E_1 & 0 & \cdots & 0 & 0\\ 0 & A_2 & E_2 & \cdots & 0 & 0\\ 0 & 0 & A_3 & \ddots & & \vdots\\ \vdots & \vdots & & \ddots & \ddots & \vdots\\ 0 & 0 & 0 & \cdots & A_{r-1} & E_{r-1}\\ E_r & 0 & 0 & \cdots & 0 & A_r \end{bmatrix}, \tag{3.3}$$

where each A_i is nearly decomposable and each E_i has exactly one entry equal to 1 and $r\geqslant 2$. Use induction on n. If $n=1$ or 2, then (3.2) is actually an equality. Assume that (3.2) holds for all nearly decomposable $t\times t$ matrices, $t<n$. Let A_i be $n_i\times n_i, i=1,\dots,r$. Then, by Theorem 3.5, Chapter 3, and the induction hypothesis,

$$\begin{aligned}\operatorname{per}(A)=\operatorname{per}(PAQ) &\geqslant \prod_{i=1}^{r}\operatorname{per}(A_i)+1\\ &\geqslant \prod_{i=1}^{r}\left(s(A_i)-2n_i+2\right)+1.\end{aligned} \tag{3.4}$$

ISBN 0-201-13505-1

Clearly if $a_1,\ldots,a_r$ are positive integers, then

$$\prod_{i=1}^{r} a_i \geqslant \sum_{i=1}^{r} a_i-(r-1). \tag{3.5}$$

(See Problem 8.) Set $a_i=s(A_i)-2n_i+2, i=1,\ldots,n$, where obviously $s(A_i)\geqslant 2n_i$ if $n_i\geqslant 2$, and $s(A_i)=1$ if $n_i=1$. Then it follows from (3.4) and (3.5) that

$$\begin{aligned}\operatorname{per}(A) &\geqslant \sum_{i=1}^{r}(s(A_i)-2n_i+2)-r+2\\ &=\sum_{i=1}^{r} s(A_i)-2n+2r-r+2\\ &=s(A)-r-2n+2r-r+2\\ &=s(A)-2n+2.\end{aligned}$$

Now suppose that A is any fully indecomposable matrix. If A is not nearly decomposable, there must exist a positive entry $a_{i_1j_1}=1$ in A, so that $A-E_{i_1j_1}$ is a fully indecomposable (0,1)-matrix. If $A-E_{i_1j_1}$ is not nearly decomposable, then there exists a positive entry $a_{i_2j_2}=1$ in $A-E_{i_1j_1}$ so that $A-E_{i_1j_1}-E_{i_2j_2}$ is fully indecomposable; and so on. Thus we must finally obtain a nearly decomposable matrix B satisfying

$$A=B+\sum_{t=1}^{m} E_{i_tj_t}.$$

By Theorem 3.7, Chapter 3,

$$\operatorname{per}(A)\geqslant \operatorname{per}(B)+m,$$

and hence, applying inequality (3.2) to the nearly decomposable matrix B, we have

$$\operatorname{per}(A)\geqslant s(B)-2n+2+m.$$

But $s(B)+m=s(A)$, and thus

$$\operatorname{per}(A)\geqslant s(A)-2n+2.$$

This concludes the proof of Theorem 3.2. ■

A necessary and sufficient condition for equality in (3.2) was obtained by Brualdi and Gibson [283].

Proof of Theorem 3.1 (continued). We proceed to establish the condition for equality in (3.1). We can assume without loss of generality that

ISBN 0-201-13505-1

$r_1 \geqslant r_2 \geqslant \cdots \geqslant r_n$, and we prove that

$$\operatorname{per}(A) = r_1 \tag{3.6}$$

if and only if $r_2 = \cdots = r_n = 2$.

Now by Theorem 3.2,

$$\begin{aligned}\operatorname{per}(A) &\geqslant s(A) - 2n + 2\\ &= r_1 + \sum_{i=2}^{n} (r_i - 2),\end{aligned}$$

and therefore, if (3.6) holds, then

$$\begin{aligned} r_1 &= \operatorname{per}(A)\\ &\geqslant r_1 + \sum_{i=2}^{n} (r_i - 2),\end{aligned}$$

and since $r_i - 2 \geqslant 0$ for all i, we must have

$$r_2 = \cdots = r_n = 2.$$

Conversely, let A be a fully indecomposable (0, 1)-matrix with $r_1 \geqslant r_2 = \cdots = r_n = 2$. We have to prove that this implies (3.6)—that is, that $\operatorname{per}(A(1|j)) = 1$ whenever $a_{1j} = 1$. We shall, in fact, establish a somewhat stronger conclusion:

$$\operatorname{per}(A(1|j)) = 1, \tag{3.7}$$

$j = 1, \ldots, n$. We assert that it suffices to prove (3.7) for nearly decomposable matrices. For, suppose that A is fully indecomposable but not nearly decomposable. Then there must exist an entry $a_{1j_1} = 1$ in the first row of A such that $A - E_{1j_1}$ is a fully indecomposable (0, 1)-matrix. Again, if $A - E_{1j_1}$ is not nearly decomposable, then there is an entry $a_{1j_2} = 1$ in the first row such that $A - E_{1j_1} - E_{1j_2}$ is fully indecomposable; and so on. Thus we must finally obtain a nearly decomposable (0, 1)-matrix

$$B = A - \sum_{t=1}^{m} E_{1j_t}$$

with row sums $r_1 - m \geqslant r_2 = \cdots = r_n = 2$. Now,

$$B(1|j) = A(1|j),$$

$j = 1, \ldots, n$, and therefore condition (3.6) is equivalent to

$$\operatorname{per}(B(1|j)) = 1,$$

ISBN 0-201-13505-1

$j=1,\ldots,n$. Hence we can assume without loss of generality that A is nearly decomposable with row sums $r_1 \geqslant r_2 = \cdots = r_n = 2$. Let Q and R be permutation matrices such that QAR is in the form (3.3), where $r \geqslant 2$, A_1 is an n_1-square nearly decomposable matrix, $A_i = 1$ for $i=2,\ldots,n$, and each E_i has exactly one positive entry. (See Hartfiel's refinement of the Sinkhorn–Knopp theorem in Section 3.3.) Let Q and R be so chosen that the positive entries of E_1 and E_r lie in the $(1, n_1+1)$ and the $(n,1)$ positions, respectively. First note that, if $r_1 = 2$—that is, $n_1 = 1$—then QAR is the sum of the identity matrix I_n and the permutation matrix P_n with 1's in the superdiagonal. It is easy to verify that in this case the permanent of every $(n-1)\times(n-1)$ submatrix of $I_n + P_n$ is 1. We prove the case $r_1 > 2$ by induction on n. Clearly the first row sum of QAR is r_1, and therefore the first row sum of A_1 is $r_1 - 1$, and its other row sums are 2. Hence by the induction hypothesis,

$$\operatorname{per}(A_1(1|j)) = 1, \tag{3.8}$$

$j=1,\ldots,n_1$. It is easy to see from the structure of the matrix QAR (even without any assumption on the form of A_1) that

$$\operatorname{per}((QAR)(1|j)) = \operatorname{per}(A_1(1|j)),$$

$j=1,\ldots,n_1$, and

$$\operatorname{per}((QAR)(1|j)) = \operatorname{per}(A_1(1|1)),$$

$j = n_1+1,\ldots,n$. The result now follows by (3.8). ■

Corollary [236]. *If A is a fully indecomposable* (0, 1)-*matrix with maximal eigenvalue $r(A)$, then*

$$\operatorname{per}(A) \geqslant r(A),$$

with equality if and only if all the row sums of A are 2.

The corollary is an immediate consequence of the inequality in Theorem 3.1 and a classical result of Frobenius which states that $r(A) \leqslant \max_i r_i$, where equality can hold if and only if all the row sums are equal.

Gibson [206] used Hall's inequality (Theorem 1.2) to obtain the following improvement of the result in Theorem 3.2.

Theorem 3.3. *Let A be a fully indecomposable $n \times n$ matrix with row sums $r_1,\ldots,r_n$. If $r_i \geqslant t > 1, i=1,\ldots,n$, then*

$$\operatorname{per}(A) \geqslant s(A) - 2n + 2 + \sum_{m=1}^{t-1} (m! - 1).$$

ISBN 0-201-13505-1

4.4 Nonnegative Matrices

In this section we discuss lower bounds for the permanents of general nonnegative matrices.

If an $n\times n$ matrix $A=(a_{ij})$ happens to be positive and $a=\min_{i,j}a_{ij}>0$, then clearly

$$\operatorname{per}(A)\geqslant n!a^n, \tag{4.1}$$

where equality can hold if and only if $a_{ij}=a$ for all i and j. The bound in (4.1) can be improved if further information is available. For example, it is easily seen that

$$\operatorname{per}(A)\geqslant n!\prod_{i=1}^{n} a_{in}^*, \tag{4.2}$$

with equality if and only if each row is a multiple of $(1,1,\ldots,1)$. (Recall the notation introduced in Section 4.2: If $A_{(i)}=(a_{i1},\ldots,a_{in})$ is the ith row of A, then $(a_{i1}^*,\ldots,a_{in}^*)$ denotes the n-tuple $A_{(i)}$ arranged in a nonincreasing order and $(a_{i1}',\ldots,a_{in}')$ the same n-tuple arranged in nondecreasing order. Thus, a_{in}^* is the least element in the ith row.)

The following result due to Jurkat and Ryser [112] gives a bound substantially better than the one in (4.2). The proof given here and the discussion of the case of equality are due to Minc [161]. In the next two theorems we shall give both lower and upper bounds, since in each case the proofs for both bounds are almost identical.

THEOREM 4.1. *Let $A=(a_{ij})$ be a nonnegative n-square matrix. Then*

$$\prod_{i=1}^{n}\sum_{t=1}^{i} a_{it}'\leqslant \operatorname{per}(A)\leqslant \prod_{i=1}^{n}\sum_{t=1}^{i} a_{it}^*. \tag{4.3}$$

If A is positive, then equality can occur in (4.3) *if and only if the first $n-1$ rows of A are multiples of* $(1,1,\ldots,1)$.

The theorem states that per(A) cannot be exceeded by the product of the least entry in the first row of A and the sum of the two smallest entries in the second row and the sum of the three smallest entries in the third row, etc. Similarly for the upper bound. Clearly an analogous result holds for the columns of A.

Proof. We prove the lower bound in (4.3) by induction on n. It is easily seen that (4.3) and the assertion about the equality hold for $n=2$. Let $n\geqslant 3$, and assume that the result holds for all $(n-1)$-square matrices. Expand the

ISBN 0-201-13505-1

permanent of A by its last row, and apply the induction hypothesis:

$$\begin{aligned}\operatorname{per}(A) &= \sum_{j=1}^{n} a_{nj} \operatorname{per}(A(n|j)) \\ &\geqslant \sum_{j=1}^{n} a_{nj} \prod_{i=1}^{n-1} \sum_{t=1}^{i} (A(n|j))'_{it}. \end{aligned} \tag{4.4}$$

Obviously

$$\sum_{t=1}^{i} (A(n|j))'_{it} \geqslant \sum_{t=1}^{i} a'_{it}, \tag{4.5}$$

$i=1,\ldots,n-1$, $j=1,\ldots,n$, and therefore

$$\begin{aligned}\operatorname{per}(A) &\geqslant \sum_{j=1}^{n} a_{nj} \prod_{i=1}^{n-1} \sum_{t=1}^{i} a'_{it} \\ &= \left(\prod_{i=1}^{n-1} \sum_{t=1}^{i} a'_{it}\right)\left(\sum_{j=1}^{n} a_{nj}\right) \\ &= \prod_{i=1}^{n} \sum_{t=1}^{i} a'_{it}. \end{aligned} \tag{4.6}$$

The upper bound is proved similarly.

The condition for equality is obviously sufficient for both inequalities in (4.3). (See Problem 10.) Now suppose that A is positive and the left-hand side of (4.3) is an equality. This can occur only if equality holds both in (4.4) and in (4.5). Now, by the induction hypothesis, (4.4) is an equality only if each of the first $n-2$ rows of $A(n|j)$ is a multiple of the $(n-1)$-tuple $(1,1,\ldots,1)$. Since $n \geqslant 3$, this condition implies that

$$a_{i1} = a_{i2} = \cdots = a_{in} \tag{4.7}$$

for $i=1,\ldots,n-2$. Next, suppose that equality holds in (4.5). Then in particular,

$$\sum_{t=1}^{n-1} (A(n|j))'_{n-1,t} = \sum_{t=1}^{n-1} a'_{n-1,t},$$

$j=1,\ldots,n$; that is,

$$\left(\sum_{t=1}^{n} a_{n-1,t}\right) - a_{n-1,j} = \sum_{t=1}^{n-1} a'_{n-1,t}$$

ISBN 0-201-13505-1

for $j=1,\ldots,n$. Hence

$$a_{n-1,1}=a_{n-1,2}=\cdots=a_{n-1,n}, \tag{4.8}$$

and the result follows by (4.7) and (4.8).

The proof of the necessity of the condition for equality on the right-hand side of (4.3) is almost identical to the above. ■

Since per(A) is invariant under a permutation of the rows of A, each of the bounds in (4.3) can be improved by a judicious permutation of the rows of A.

COROLLARY 1. *Let $A=(a_{ij})$ be a nonnegative $n\times n$ matrix. Then*

$$\max_{\sigma\in S_n}\prod_{i=1}^{n}\sum_{t=1}^{i}a'_{\sigma(i),t}\leqslant \operatorname{per}(A)\leqslant \min_{\sigma\in S_n}\prod_{i=1}^{n}\sum_{t=1}^{i}a^*_{\sigma(i),t}. \tag{4.9}$$

If A is positive, then equality can occur in (4.9) *if and only if $n-1$ of the rows of A are multiples of* $(1,1,\ldots,1)$.

Note that the condition for equality in Theorem 4.1 (or in the corollary) may not be necessary if A is merely nonnegative. For example, if

$$A=\begin{bmatrix}1&1&1&1\\1&1&1&0\\1&1&1&0\\2&0&0&0\end{bmatrix}$$

then the lower bound given by Theorem 4.1 is

$$1\times(0+1)\times(0+1+1)\times(0+0+0+2)=4$$

which is equal to per(A). Similarly, if

$$B=\begin{bmatrix}1&1&1&1\\1&1&1&1\\0&0&1&2\\2&1&0&0\end{bmatrix},$$

then the upper bound given by Theorem 4.1 is

$$1\times(1+1)\times(2+1+0)\times(2+1+0+0)=18$$

ISBN 0-201-13505-1

which is equal to per(B). Of course, neither A nor B is in the form required by the condition of equality in Theorem 4.1.

Each of the bounds in Theorem 4.1 depends on a single entry in the first row of A. In the following theorem we improve these bounds by introducing correction terms depending also on the first row sum of the matrix.

THEOREM 4.2 [161]. *Let $A=(a_{ij})$ be a nonnegative $n \times n$ matrix. Then*

$$\operatorname{per}(A) \geqslant \prod_{i=1}^{n} \sum_{t=1}^{i} a'_{it} + (r_1 - na'_{11}) \prod_{j=2}^{n} \sum_{s=1}^{j-1} a'_{js}, \tag{4.10}$$

and

$$\operatorname{per}(A) \leqslant \prod_{i=1}^{n} \sum_{t=1}^{i} a^*_{it} - (na^*_{11} - r_1) \prod_{j=2}^{n} \sum_{s=1}^{j-1} a'_{js}, \tag{4.11}$$

where r_1 denotes the first row sum of A, and a'_{11} and a^*_{11} are the least and largest entries in the first row of A, respectively.

Proof. We first prove (4.10). Let $B=(b_{ij})$ and $C=(c_{ij})$ be $n \times n$ matrices defined by

$$\begin{cases} b_{1j} = a_{1j} - a'_{11}, & \text{for } j=1,\ldots,n, \text{ and} \\ b_{ij} = a_{ij}, & \text{otherwise;} \end{cases}$$

$$\begin{cases} c_{1j} = a'_{11}, & \text{for } j=1,\ldots,n, \text{ and} \\ c_{ij} = a_{ij}, & \text{otherwise.} \end{cases}$$

Then

$$\operatorname{per}(A) = \operatorname{per}(B) + \operatorname{per}(C). \tag{4.12}$$

Now by Theorem 4.1,

$$\begin{aligned} \operatorname{per}(C) &\geqslant \prod_{i=1}^{n} \sum_{t=1}^{i} c'_{it} \\ &= \prod_{i=1}^{n} \sum_{t=1}^{i} a'_{it}. \end{aligned} \tag{4.13}$$

Expanding the permanent of B by the first row and applying Theorem 4.1

ISBN 0-201-13505-1

to the $B(1|t)$ we obtain

$$\begin{aligned}
\operatorname{per}(B) &= \sum_{t=1}^{n} b_{1t}\operatorname{per}(B(1|t)) \\
&= \sum_{t=1}^{n} (a_{1t}-a'_{11})\operatorname{per}(A(1|t)) \\
&\geqslant \sum_{t=1}^{n} (a_{1t}-a'_{11}) \prod_{j=1}^{n-1} \sum_{s=1}^{j} (A(1|t))'_{js} \\
&\geqslant \sum_{t=1}^{n} (a_{1t}-a'_{11}) \prod_{j=2}^{n} \sum_{s=1}^{j-1} a'_{js} \\
&= (r_1 - na'_{11}) \prod_{j=2}^{n} \sum_{s=1}^{j-1} a'_{js}. \qquad (4.14)
\end{aligned}$$

Inequality (4.10) now follows from (4.12), (4.13), and (4.14).

The bound in (4.11) is proved similarly using induction on n. (See Problem 14.) ■

Corollary 2. *Let $A=(a_{ij})$ be a nonnegative $n\times n$ matrix with row sums $r_1,\ldots,r_n$. Then*

$$\operatorname{per}(A) \geqslant \max_{\sigma\in S_n}\left(\prod_{i=1}^{n}\sum_{t=1}^{i} a'_{\sigma(i),t} + (r_{\sigma(1)} - na'_{\sigma(1),1})\prod_{j=2}^{n}\sum_{s=1}^{j-1} a'_{\sigma(j),s}\right),$$

and

$$\operatorname{per}(A) \leqslant \min_{\sigma\in S_n}\left(\prod_{i=1}^{n}\sum_{t=1}^{n} a^*_{\sigma(i),t} - (na^*_{\sigma(1),1} - r_{\sigma(1)})\prod_{j=2}^{n}\sum_{s=1}^{j-1} a'_{\sigma(j),s}\right).$$

4.5 Positive Semi-definite Hermitian Matrices

In this section we state and prove inequalities for permanents of positive semi-definite hermitian matrices whose entries are not necessarily real. These results were obtained by Marcus and Minc [102].

Theorem 5.1. *If $A=(a_{ij})$ is a positive semi-definite hermitian $n\times n$ matrix with row sums $r_1,\ldots,r_n$, then*

$$\operatorname{per}(A) \geqslant \frac{n!}{s(A)^n}\prod_{i=1}^{n}|r_i|^2, \qquad (5.1)$$

ISBN 0-201-13505-1

provided $s(A)=\sum_{i=1}^{n} r_i \neq 0$. *Equality holds in* (5.1) *if and only if either* A *has a zero row or the rank of* A *is* 1.

Proof. Since A is positive semi-definite hermitian, it is a Gram matrix based on some vectors $x_1,\ldots,x_n$—that is, $a_{ij}=(x_i,x_j)$, $i=1,\ldots,n$, $j=1,\ldots,n$. From Theorem 2.2, Chapter 2, we have

$$\frac{1}{n!}\operatorname{per}(A)=\|x_1*\cdots*x_n\|^2. \tag{5.2}$$

Let $u=\sum_{i=1}^{n} x_i/\sqrt{s(A)}$. Then

$$\begin{aligned}(u,u)&=\sum_{i,j}\frac{(x_i,x_j)}{s(A)}\\&=\sum_{i,j}\frac{a_{ij}}{s(A)}\\&=1;\end{aligned}$$

that is, u is a unit vector. Let $v=u*\cdots*u$, and note that by Theorem 2.3, Chapter 2, v is nonzero. Thus by the Cauchy–Schwarz inequality,

$$|(x_1*\cdots*x_n,v)|^2\leqslant\|x_1*\cdots*x_n\|^2\|v\|^2, \tag{5.3}$$

and therefore from (5.2),

$$\begin{aligned}\frac{1}{n!}\operatorname{per}(A)&\geqslant\frac{1}{\|v\|^2}|(x_1*\cdots*x_n,u*\cdots*u)|^2\\&=\frac{1}{\|v\|^2}\left|\frac{1}{n!}\operatorname{per}((x_i,u))\right|^2.\end{aligned}$$

But

$$(x_i,u)=\left(x_i,\sum_{j=1}^{n}\frac{x_j}{\sqrt{s(A)}}\right)=\sum_{j=1}^{n}\frac{(x_i,x_j)}{\sqrt{s(A)}}=\frac{r_i}{\sqrt{s(A)}},$$

$i=1,\ldots,n$. Hence

$$\begin{aligned}\frac{1}{n!}\operatorname{per}(A)&\geqslant\frac{1}{\|v\|^2}\left|\frac{1}{n!}n!\prod_{i=1}^{n}\frac{r_i}{\sqrt{s(A)}}\right|^2\\&=\frac{1}{\|v\|^2}\frac{1}{s(A)^n}\prod_{i=1}^{n}|r_i|^2.\end{aligned} \tag{5.4}$$

ISBN 0-201-13505-1

Now,

$$\begin{aligned}\|v\|^2 &= \|u*\cdots*u\|^2\\ &= \frac{1}{n!}\operatorname{per}((u,u))\\ &= \frac{1}{n!}n!\\ &= 1,\end{aligned} \tag{5.5}$$

and therefore, combining (5.4) and (5.5), we have the inequality (5.1).

For equality to hold in (5.1), the inequality (5.3), which is an application of the Cauchy–Schwarz inequality, must be an equality, and thus $x_1*\cdots*x_n$ and $u*\cdots*u$ must be linearly dependent. Since $u\neq 0$, it follows from Theorem 2.3, Chapter 2, that either (i) $x_i=0$ for some i, or (ii) $x_i=c_iu$ for some nonzero scalars $c_i, i=1,\ldots,n$. If (i) holds, then clearly A has a zero row, and if (ii) holds, the rank of A is 1. Conversely, if A has a zero row, then obviously both sides of (5.1) vanish. If A has rank 1, then

$$a_{ij}=c_i\bar{c}_j,$$

$i,j=1,\ldots,n$, and

$$\operatorname{per}(A)=n!\prod_{i=1}^{n}|c_i|^2.$$

Now,

$$r_i=c_i\sum_{j=1}^{n}\bar{c}_j,$$

and therefore

$$\prod_{i=1}^{n}|r_i|^2=\prod_{i=1}^{n}|c_i|^2\left|\sum_{j=1}^{n}c_j\right|^{2n}$$

and

$$s(A)^n=\left(\sum_{i=1}^{n}r_i\right)^n=\left(\sum_{i=1}^{n}c_i\sum_{j=1}^{n}\bar{c}_j\right)^n=\left|\sum_{j=1}^{n}c_j\right|^{2n}.$$

Hence

$$\frac{n!}{s(A)^n}\prod_{i=1}^{n}|r_i|^2=n!\prod_{i=1}^{n}|c_i|^2=\operatorname{per}(A),$$

completing the proof. ■

ISBN 0-201-13505-1

Corollary 1. *If A is a doubly stochastic positive semi-definite hermitian $n \times n$ matrix, then*

$$\operatorname{per}(A) \geqslant n!/n^n. \tag{5.6}$$

Equality holds in (5.6) *if and only if $A = J_n$, the $n \times n$ matrix all of whose entries are $1/n$.*

The condition for equality is based on the fact that J_n is the only doubly stochastic $n \times n$ matrix of rank 1.

Theorem 5.2. *If A is positive semi-definite hermitian with eigenvalues $\lambda_1, \ldots, \lambda_n$, then*

$$\operatorname{per}(A) \geqslant n! \sum_{t=1}^{n} |\xi_t|^2 \lambda_t^n, \tag{5.7}$$

where ξ_t is the product of the coordinates of the unit eigenvector corresponding to λ_t.

Proof. Let $A = U^* D U$, where U is unitary and $D = \operatorname{diag}(\lambda_1, \ldots, \lambda_n)$. Using the Binet–Cauchy theorem (Theorem 1.3, Chapter 2) we compute:

$$\begin{aligned}\operatorname{per}(A) &= \sum_{\omega \in G_{n,n}} \frac{1}{\mu(\omega)} \operatorname{per}(U^*[1, \ldots, n|\omega]) \operatorname{per}((DU)[\omega|1, \ldots, n]) \\ &= \sum_{\omega \in G_{n,n}} \frac{1}{\mu(\omega)} |\operatorname{per}(U[\omega|1, \ldots, n])|^2 \prod_{t=1}^{n} \lambda_t^{m_t(\omega)}. \end{aligned} \tag{5.8}$$

If we retain only the sequences $(t, \ldots, t)$, $t = 1, \ldots, n$, and discard the other ω in $G_{n,n}$, then (5.8) gives

$$\begin{aligned}\operatorname{per}(A) &\geqslant \sum_{t=1}^{n} \frac{1}{n!} |\operatorname{per}(U[t, \ldots, t|1, \ldots, n])|^2 \lambda_t^n \\ &= \frac{1}{n!} \sum_{t=1}^{n} \left| n! \prod_{j=1}^{n} u_{tj} \right|^2 \lambda_t^n \\ &= n! \sum_{t=1}^{n} |\xi_t|^2 \lambda_t^n. \end{aligned}$$

■

Corollary 2. *If all the row sums of a positive semi-definite hermitian $n \times n$ matrix A are equal to k, then*

$$\operatorname{per}(A) \geqslant n! \left(\frac{k}{n} \right)^n. \tag{5.9}$$

ISBN 0-201-13505-1

The bound in Theorem 5.2 can be improved by retaining an additional term in (5.8). Our next result in a way combines the bound in Theorem 5.2 with that in Theorem 2.5, Chapter 2.

THEOREM 5.3. *If U is a unitary matrix, and $A = U^*\operatorname{diag}(\lambda_1,\ldots,\lambda_n)U$, where the $\lambda_1,\ldots,\lambda_n$ are nonnegative, then*

$$\operatorname{per}(A) \geqslant n! \sum_{t=1}^{n} |\xi_t|^2 \lambda_t^n + |\operatorname{per}(U)|^2 \det(A), \tag{5.10}$$

where the ξ_t are as defined in Theorem 5.2.

Proof. In addition to the sequences $(t,\ldots,t), t=1,\ldots,n$, we retain on the right-hand side of (5.8) the sequence $(1,\ldots,n)$; the other sequences of $G_{n,n}$ are discarded as in the proof of Theorem 5.2. Then (5.8) yields

$$\operatorname{per}(A) \geqslant n! \sum_{t=1}^{n} |\xi_t|^2 \lambda_t^n + |\operatorname{per}(U)|^2 \prod_{i=1}^{n} \lambda_i,$$

and the result follows. ■

Example 5.1. Evaluate the bounds in (5.1), (5.7), and (5.10) for the permanents of

$$A = \begin{bmatrix} \frac{3}{5} & \frac{4}{5} \\ \frac{4}{5} & -\frac{3}{5} \end{bmatrix} \begin{bmatrix} 50 & 0 \\ 0 & 25 \end{bmatrix} \begin{bmatrix} \frac{3}{5} & \frac{4}{5} \\ \frac{4}{5} & -\frac{3}{5} \end{bmatrix} = \begin{bmatrix} 34 & 12 \\ 12 & 41 \end{bmatrix}$$

and

$$B = \begin{bmatrix} \frac{2}{3} & \frac{2}{3} & -\frac{1}{3} \\ \frac{1}{3} & -\frac{2}{3} & -\frac{2}{3} \\ \frac{2}{3} & -\frac{1}{3} & \frac{2}{3} \end{bmatrix} \begin{bmatrix} 9 & 0 & 0 \\ 0 & 6 & 0 \\ 0 & 0 & 3 \end{bmatrix} \begin{bmatrix} \frac{2}{3} & \frac{1}{3} & \frac{2}{3} \\ \frac{2}{3} & -\frac{2}{3} & -\frac{1}{3} \\ -\frac{1}{3} & -\frac{2}{3} & \frac{2}{3} \end{bmatrix} = \begin{bmatrix} 7 & 0 & 2 \\ 0 & 5 & 2 \\ 2 & 2 & 6 \end{bmatrix}.$$

The bounds for $\operatorname{per}(A)$ are

$$\frac{2!}{99^2} \times 46^2 \times 53^2 = 1213 \text{ (approx.)},$$

$$2\left(\left(\tfrac{3}{5}\times\tfrac{4}{5}\right)^2 50^2 + \left(\tfrac{4}{5}\times\tfrac{3}{5}\right)^2 25^2\right) = 1440,$$

and

$$1440 + \left(\tfrac{7}{25}\right)^2 \times 1250 = 1538.$$

In fact, $\operatorname{per}(A) = 1538$.

Similarly we compute the three bounds for $\operatorname{per}(B)$ and obtain: 136 (approx.), 128, and 130. Actually the permanent of B is equal to 258.

ISBN 0-201-13505-1

Problems

1. Is the condition $\mathrm{Per}(A)>0$ in Theorem 1.2 (case $t \leqslant m$) superfluous?
2. Show that the bounds in Theorem 1.2 are the best possible, in general.
3. Let

$$A=\begin{bmatrix} 1 & 1 & 1 & 1 & 1 & 1 \\ 1 & 1 & 1 & 1 & 0 & 1 \\ 1 & 1 & 0 & 1 & 1 & 0 \\ 1 & 1 & 1 & 0 & 0 & 1 \\ 1 & 1 & 0 & 1 & 1 & 0 \\ 1 & 1 & 1 & 1 & 1 & 0 \end{bmatrix}.$$

   Use Theorem 2.1 to obtain a lower bound for $\mathrm{per}(A)$.
4. Let A be the matrix in Problem 3. Use formula (2.7) to improve the bound in Problem 3.
5. Let A be the matrix in Problem 3. Obtain a lower bound for $\mathrm{per}(A)$ by applying Theorem 2.1 and formula (2.7) to columns of A.
6. Let

$$B=\begin{bmatrix} 1 & 1 & 0 & 1 & 1 & 1 \\ 1 & 1 & 1 & 0 & 1 & 0 \\ 1 & 0 & 1 & 0 & 0 & 1 \\ 0 & 0 & 1 & 1 & 1 & 0 \\ 0 & 1 & 0 & 1 & 1 & 1 \\ 1 & 1 & 0 & 1 & 0 & 0 \end{bmatrix}.$$

   Use Theorem 3.2 to obtain a lower bound for $\mathrm{per}(B)$.
7. Let B be the matrix in Problem 6. Use Theorem 3.3 to improve the bound in Problem 6. What lower bound is obtained by means of Theorem 2.1?
8. Prove inequality (3.5).
9. Let $A=(a_{ij})$ be an $n\times n$ nonnegative matrix, and let $a=\min_{i,\,j} a_{ij}$. Suppose that A contains a $k\times k$ submatrix, $k<n$, whose permanent is equal to p. What is the least possible value for $\mathrm{per}(A)$?
10. Show that, if every row of an $n\times n$ matrix $A=(a_{ij})$ is a multiple of $(1,1,\ldots,1)$, then the left inequality in (4.3) becomes an equality; that is,

$$\mathrm{per}(A)=\prod_{i=1}^{n}\sum_{t=1}^{i} a'_{it}.$$

11. Use Theorem 4.1 to find a lower bound for the permanent of

$$A=\begin{bmatrix} 2 & 1 & 1 & 2 \\ 3 & 1 & 3 & 1 \\ 2 & 2 & 2 & 2 \\ 2 & 2 & 1 & 2 \end{bmatrix}.$$

ISBN 0-201-13505-1

12. Use Corollary 1 to Theorem 4.1 to find a lower bound for the permanent of the matrix A in Problem 11.
13. Repeat the exercise in Problems 11 and 12 using the result in Theorem 4.2.
14. Prove the inequality (4.11).
15. Let $A = U^*DU$, where U and D are a unitary and a nonnegative diagonal matrix, respectively. Use the method in the proof of Theorem 5.2 to show that

$$\operatorname{per}(A) \geqslant |\operatorname{per}(U)|^2 \det(A).$$

(*Note*: This result is weaker than that in Theorem 2.5, Chapter 2, since by Corollary 2 to Theorem 2.4, Chapter 2, $|\operatorname{per}(U)| \leqslant 1$.)
16. Deduce from Theorem 5.2 that, if A is a positive definite symmetric doubly stochastic $n \times n$ matrix, then

$$\operatorname{per}(A) \geqslant n!/n^n.$$

17. Explain why in Example 5.1 the bound in (5.10) gave the exact value for per(A).
18. Let Λ_n^k denote the set of $n \times n$ (0, 1)-matrices with exactly k 1's in each row and column. Use Theorem 3.2 to find a lower bound for per(A), $A \in \Lambda_n^k$.
19. Prove that, if $A \in \Lambda_n^k$, then

$$\operatorname{per}(A) \geqslant (n-1)(k-2) + \sum_{m=0}^{k-1} m!.$$

ISBN 0-201-13505-1

CHAPTER 5

The van der Waerden Conjecture

5.1 The Marcus–Newman Theory

The main cause of the recent revival of interest in permanents was the appearance in 1959 of the classical paper of Marcus and Newman [62] on the van der Waerden conjecture. Recall that the conjecture asserts that, if S is a doubly stochastic $n \times n$ matrix, then

$$\operatorname{per}(S) \geqslant n!/n^n, \tag{1.1}$$

and that equality can hold in (1.1) if and only if S is J_n, the matrix all of whose entries equal $1/n$.

In spite of many efforts, the van der Waerden conjecture remains unresolved. In this chapter we describe its current status.

Marcus and Newman hoped to prove the van der Waerden conjecture by establishing properties of matrices whose permanents are minimal in Ω_n and then showing that only the matrix J_n satisfies these properties. In spite of the failure of this program, the work of Marcus and Newman represents the major part of the progress achieved so far in the area of the van der Waerden conjecture. In this section we present their most important results.

Call a doubly stochastic $n \times n$ matrix satisfying

$$\operatorname{per}(A) = \min_{S \in \Omega_n} \operatorname{per}(S) \tag{1.2}$$

a *minimizing matrix.*

THEOREM 1.1 [62]. *A minimizing matrix is fully indecomposable.*

ENCYCLOPEDIA OF MATHEMATICS and Its Applications, Gian-Carlo Rota (ed.). Vol. 6: Henryk Minc, Permanents

ISBN 0-201-13505-1

Proof. Let $A \in \Omega_n$ be a minimizing matrix. Suppose that A is partly decomposable. Then, by Theorem 3.1, Chapter 3, there exist permutation matrices P and Q such that $PAQ = B \dotplus C$, where $B \in \Omega_k$, $C \in \Omega_{n-k}$. We shall show that there exists a doubly stochastic matrix whose permanent is less than per(A). Since per(A)>0, we can assume without loss of generality that $b_{kk}\operatorname{per}((PAQ)(k|k))>0$ and $c_{11}\operatorname{per}((PAQ)(k+1|k+1))>0$. Let ε be any positive number smaller than $\min(b_{kk}, c_{11})$ and let

$$G(\varepsilon) = PAQ - \varepsilon(E_{kk} + E_{k+1,k+1}) + \varepsilon(E_{k,k+1} + E_{k+1,k}).$$

Then $G(\varepsilon) \in \Omega_n$, and

$$\begin{aligned}\operatorname{per}(G(\varepsilon)) = &\operatorname{per}(PAQ) - \varepsilon\operatorname{per}((PAQ)(k|k)) + \varepsilon\operatorname{per}((PAQ)(k|k+1)) \\ &- \varepsilon\operatorname{per}((PAQ)(k+1|k+1)) + \varepsilon\operatorname{per}((PAQ)(k+1|k)) + O(\varepsilon^2) \\ = &\operatorname{per}(A) - \varepsilon\big(\operatorname{per}((PAQ)(k|k)) + \operatorname{per}((PAQ)(k+1|k+1))\big) + O(\varepsilon^2),\end{aligned}$$

since $\operatorname{per}((PAQ)(k|k+1)) = \operatorname{per}((PAQ)(k+1|k)) = 0$, by the Frobenius–König theorem. Also, $\operatorname{per}((PAQ)(k|k)) + \operatorname{per}((PAQ)(k+1|k+1)) > 0$, and therefore, for sufficiently small positive ε,

$$\operatorname{per}(G(\varepsilon)) < \operatorname{per}(A),$$

contradicting the assumption that A is a minimizing matrix. ■

In our next theorem we prove a fundamental property of a minimizing matrix.

THEOREM 1.2 [62]. *If* $A = (a_{ij})$ *is a minimizing matrix, then* $a_{hk} > 0$ *implies* $\operatorname{per}(A(h|k)) = \operatorname{per}(A)$.

Proof. Let $C(A)$ be the face of Ω_n of least dimension containing A in its interior. In other words,

$$C(A) = \{X = (x_{ij}) \in \Omega_n | x_{ij} = 0 \quad \text{if } (i,j) \in Z\},$$

where $Z = \{(i,j) | a_{ij} = 0\}$. Then $C(A)$ is defined by the following conditions:

$$\sum_{j=1}^{n} x_{ij} = 1, \qquad i = 1, \ldots, n;$$

$$\sum_{i=1}^{n} x_{ij} = 1, \qquad j = 1, \ldots, n;$$

$$x_{ij} \geqslant 0, \qquad i,j = 1, \ldots, n;$$

$$x_{ij} = 0, \qquad (i,j) \in Z.$$

ISBN 0-201-13505-1

Now A is an absolute minimum for the permanent function over the whole Ω_n, and, since A is in the interior of $C(A)$, it must be a relative minimum for the function on $C(A)$. Hence we may introduce Lagrange multipliers and set up the function on $C(A)$:

$$F(X)=\operatorname{per}(X)-\sum_{i=1}^{n}\lambda_i\left(\sum_{k=1}^{n}x_{ik}-1\right)-\sum_{j=1}^{n}\mu_j\left(\sum_{k=1}^{n}x_{kj}-1\right). \tag{1.3}$$

Now for $(i,j)\notin Z$,

$$\partial F(X)/\partial x_{ij}=\operatorname{per}(X(i|j))-\lambda_i-\mu_j.$$

Therefore

$$\operatorname{per}(A(i|j))=\lambda_i+\mu_j, \tag{1.4}$$

and thus

$$\begin{aligned}\operatorname{per}(A)&=\sum_{j=1}^{n}a_{ij}\operatorname{per}(A(i|j))\\&=\sum_{j=1}^{n}a_{ij}(\lambda_i+\mu_j)\\&=\lambda_i+\sum_{j=1}^{n}a_{ij}\mu_j,\end{aligned} \tag{1.5}$$

$i=1,\ldots,n$; and similarly

$$\begin{aligned}\operatorname{per}(A)&=\sum_{i=1}^{n}a_{ij}\operatorname{per}(A(i|j))\\&=\mu_j+\sum_{i=1}^{n}\lambda_i a_{ij},\end{aligned} \tag{1.6}$$

$j=1,\ldots,n$. Now let $\lambda=(\lambda_1,\ldots,\lambda_n)$, $\mu=(\mu_1,\ldots,\mu_n)$, $e=(1,\ldots,1)$. Then from (1.5) and (1.6) we have

$$\operatorname{per}(A)e=\lambda+A\mu, \tag{1.7}$$

$$\operatorname{per}(A)e=A^{\mathrm{T}}\lambda+\mu. \tag{1.8}$$

Premultiply (1.7) by A^{T}:

$$\operatorname{per}(A)e=A^{\mathrm{T}}\lambda+A^{\mathrm{T}}A\mu, \tag{1.9}$$

since $A^{\mathrm{T}}e=e$. Subtract (1.8) from (1.9):

$$A^{\mathrm{T}}A\mu=\mu. \tag{1.10}$$

Similarly,

$$AA^{\mathrm{T}}\lambda=\lambda. \tag{1.11}$$

Now, $A^{\mathrm{T}}A$ and AA^{T} are both fully indecomposable (see Problem 11, Chapter 3), and 1 is a simple eigenvalue of each of them. Therefore both λ and μ are multiples of e—say, $\lambda=ce$ and $\mu=de$—and it follows from (1.4) that

$$\operatorname{per}(A(i|j))=c+d,$$

for all $(i,j)\notin Z$. Hence

$$\begin{aligned}\operatorname{per}(A)&=\sum_{j=1}^{n}a_{ij}\operatorname{per}(A(i|j))\\&=\sum_{j=1}^{n}a_{ij}(c+d)\\&=c+d\\&=\operatorname{per}(A(i|j)),\end{aligned}$$

for all $(i,j)\notin Z$. ■

Armed with Theorem 1.2 we now proceed to prove the following major partial result on the van der Waerden conjecture, due to Marcus and Newman [62].

THEOREM 1.3. *If a positive n-square matrix A is a minimizing matrix, then*

$$\operatorname{per}(A)=\operatorname{per}(J_n)=n!/n^n.$$

Proof. By Theorem 1.2, $\operatorname{per}(A(i|j))=\operatorname{per}(A)$ for all i and j. Let $G=J_2\dotplus I_{n-2}$. Then

$$\begin{aligned}\operatorname{per}(GA)=\tfrac{1}{2}\operatorname{per}(A)&+\tfrac{1}{4}\operatorname{per}(A_{(1)},A_{(1)},A_{(3)},\dots,A_{(n)})\\&+\tfrac{1}{4}\operatorname{per}(A_{(2)},A_{(2)},A_{(3)},\dots,A_{(n)}),\end{aligned}$$

where $\operatorname{per}(v_1,v_2,\dots,v_n)$ denotes the permanent of the matrix whose rows are the n-tuples $v_1,v_2,\dots,v_n$. But, expanding $\operatorname{per}(A_{(1)},A_{(1)},A_{(3)},\dots,A_{(n)})$ by

ISBN 0-201-13505-1

the second row, we obtain

$$\begin{aligned}\operatorname{per}(A_{(1)},A_{(1)},A_{(3)},\ldots,A_{(n)}) &= \sum_{j=1}^{n} a_{1j}\operatorname{per}(A(2|j))\\ &= \operatorname{per}(A),\end{aligned}$$

since $\operatorname{per}(A(2|j))=\operatorname{per}(A(1|j))$, $j=1,\ldots,n$. Similarly we can show that $\operatorname{per}(A_{(2)},A_{(2)},A_{(3)},\ldots,A_{(n)})=\operatorname{per}(A)$. It follows that $\operatorname{per}(GA)=\operatorname{per}(A)$. Hence GA is a positive minimizing matrix, and $\operatorname{per}((GA)(i|j))=\operatorname{per}(A)$ for all i and j.

Now let $P_1,P_2,\ldots,P_{n-2}$ be n-square permutation matrices such that

$$\begin{aligned}P_1GP_1^{\mathrm{T}} &= I_1\dotplus J_2\dotplus I_{n-3}\\ P_2GP_2^{\mathrm{T}} &= I_2\dotplus J_2\dotplus I_{n-4}\\ &\vdots\\ P_{n-3}GP_{n-3}^{\mathrm{T}} &= I_{n-3}\dotplus J_2\dotplus I_1\\ P_{n-2}GP_{n-2}^{\mathrm{T}} &= I_{n-2}\dotplus J_2.\end{aligned}$$

Then it is easily seen that

$$\begin{aligned}M_n &= P_{n-2}GP_{n-2}^{\mathrm{T}}P_{n-3}GP_{n-3}^{\mathrm{T}}\cdots P_2GP_2^{\mathrm{T}}P_1GP_1^{\mathrm{T}}G\\ &=(I_{n-2}\dotplus J_2)(I_{n-3}\dotplus J_2\dotplus I_1)\cdots(I_2\dotplus J_2\dotplus I_{n-4})(I_1\dotplus J_2\dotplus I_{n-3})(J_2\dotplus I_{n-2})\\ &=\begin{bmatrix}\frac{1}{2} & \frac{1}{2} & 0 & 0 & 0 & \cdots & 0 & 0\\ \frac{1}{4} & \frac{1}{4} & \frac{1}{2} & 0 & 0 & \cdots & 0 & 0\\ \frac{1}{8} & \frac{1}{8} & \frac{1}{4} & \frac{1}{2} & 0 & \cdots & 0 & 0\\ \cdots & \cdots & \cdots & \cdots & \cdots & \cdots & \cdots & \cdots\\ \frac{1}{2^{n-1}} & \frac{1}{2^{n-1}} & \frac{1}{2^{n-2}} & \cdots & & \cdots & \frac{1}{4} & \frac{1}{2}\\ \frac{1}{2^{n-1}} & \frac{1}{2^{n-1}} & \frac{1}{2^{n-2}} & \cdots & & \cdots & \frac{1}{4} & \frac{1}{2}\end{bmatrix}.\end{aligned}\tag{1.12}$$

The matrix M_n is doubly stochastic and primitive (as a matter of fact, totally indecomposable) with maximal eigenvalue 1; the moduli of other eigenvalues are strictly less than 1. It follows that

$$L_n=\lim_{t\to\infty} M_n^t=J_n,$$

since L_n is a doubly stochastic matrix of rank 1.

ISBN 0-201-13505-1

Now we have shown that premultiplication of a positive minimizing matrix by G produces a positive minimizing matrix. Clearly, premultiplication by a permutation matrix produces the same result. It follows that M_nA is a positive minimizing matrix. Reasoning in the same manner with M_n we deduce that

$$\operatorname{per}(M_n^tA)=\operatorname{per}(A)$$

for any positive integer t. Hence

$$\begin{aligned}\operatorname{per}(A)&=\lim_{t\to\infty}\operatorname{per}(M_n^tA)\\&=\operatorname{per}\left(\left(\lim_{t\to\infty}M_n^t\right)A\right)\\&=\operatorname{per}(J_nA)\\&=\operatorname{per}J_n\\&=n!/n^n.\end{aligned}$$

■

We shall show that there can be at most one positive minimizing matrix. Before we can do that, we need a preliminary result that is of interest by itself.

THEOREM 1.4 [62]. *If A is a positive doubly stochastic matrix sufficiently close to J_n, and $A\neq J_n$, then*

$$\operatorname{per}(A)>\operatorname{per}(J_n).$$

In other words, the permanent function has a strict local minimum at J_n.

Proof. Let $L(A)$ be the line segment through J_n and A, intersecting the boundary of Ω_n in B. That is, if $a_{st}=\min_{i,j}(a_{ij})$ and $\theta_0=1-na_{st}>0$, then

$$B=\frac{1}{\theta_0}(A-(1-\theta_0)J_n)$$

is a doubly stochastic matrix with at least one zero, and

$$L(A)=\{S\,|\,S=\theta B+(1-\theta)J_n,\qquad 0\leqslant\theta\leqslant 1\}.$$

Clearly $A=\theta_0B+(1-\theta_0)J_n\in L(A)$. Define a function ρ on the interval $[0,1]$:

$$\rho(\theta)=\operatorname{per}(\theta B+(1-\theta)J_n).$$

ISBN 0-201-13505-1

By McLaurin's theorem,

$$\begin{aligned}\operatorname{per}(A)&=\rho(\theta_0)\\&=\rho(0)+\theta_0\rho'(0)+\tfrac{1}{2}\theta_0^2\rho''(0)+O\left(\theta_0^3\right)\\&=\operatorname{per}(J_n)+\theta_0\rho'(0)+\tfrac{1}{2}\theta_0^2\rho''(0)+O\left(\theta_0^3\right).\end{aligned}\tag{1.13}$$

We show that $\rho'(0)=0$ and that $\rho''(0)>0$. It will follow that $\operatorname{per}(A)>\operatorname{per}(J_n)$ for sufficiently small θ_0. We compute

$$\rho'(\theta)=\sum_{i,j}\left(b_{ij}-\frac{1}{n}\right)\operatorname{per}((\theta B+(1-\theta)J_n)(i|j)),\tag{1.14}$$

and therefore

$$\begin{aligned}\rho'(0)&=\sum_{i,j}\left(b_{ij}-\frac{1}{n}\right)\operatorname{per}(J_n(i|j))\\&=\left(\frac{(n-1)!}{n^{n-1}}\right)\sum_{i,j}\left(b_{ij}-\frac{1}{n}\right)\\&=0.\end{aligned}\tag{1.15}$$

From (1.14) we have

$$\begin{aligned}\rho''(0)&=\sum_{i,j}\left(b_{ij}-\frac{1}{n}\right)\sum_{\substack{h\neq i\\k\neq j}}\left(b_{hk}-\frac{1}{n}\right)\operatorname{per}(J_n(i,h|j,k))\\&=\frac{(n-2)!}{n^{n-2}}\sum_{i,j}\left(b_{ij}-\frac{1}{n}\right)\left(n-2+b_{ij}-\frac{(n-1)^2}{n}\right)\\&=\frac{(n-2)!}{n^{n-2}}\sum_{i,j}\left(b_{ij}-\frac{1}{n}\right)^2\\&\geqslant(n-2)!/n^n,\end{aligned}\tag{1.16}$$

since at least one of the b_{ij} is 0, and therefore $\Sigma_{i,j}(b_{ij}-1/n)^2\geqslant 1/n^2$. Hence $\rho''(0)>0$, and the result follows by (1.13) and (1.15). ■

We now exploit Theorem 1.4 to obtain a statement on the uniqueness of a positive minimizing matrix, if indeed there is one.

THEOREM 1.5 [62]. *If A is a positive minimizing matrix, then $A=J_n$.*

ISBN 0-201-13505-1

Proof. Let M_n be the matrix (1.12), and consider the sequence of matrices

$$A, M_n A, M_n^2 A, \dots. \tag{1.17}$$

We show that, if $A \neq J_n$, then all terms of the sequence (1.17) are different from J_n. Suppose that $A \neq J_n$ and $M_n^h A = J_n$. Let s be the smallest positive integer for which $M_n^s A = J_n$. Put $M_n^{s-1} A = B$. Then $M_n B = J_n$, and thus all rows of $M_n B$ are equal to $(1/n)e$; that is,

$$\begin{aligned} \tfrac{1}{2}B_{(1)} + \tfrac{1}{2}B_{(2)} &= \tfrac{1}{4}B_{(1)} + \tfrac{1}{4}B_{(2)} + \tfrac{1}{2}B_{(3)} \\ &\vdots \\ &= \frac{1}{2^{n-1}}B_{(1)} + \frac{1}{2^{n-1}}B_{(2)} + \cdots + \frac{1}{2}B_{(n)} \\ &= \frac{1}{n}e. \end{aligned}$$

But this implies that

$$\tfrac{1}{2}(B_{(1)} + B_{(2)}) = B_{(3)} = B_{(4)} = \cdots = B_{(n)} = \frac{1}{n}e.$$

Now, B is a positive minimizing matrix, and therefore, by Theorem 1.2, all its subpermanents of order $n-1$ are equal. We compute

$$\begin{aligned} \operatorname{per}(B(1|j)) &= \sum_{\substack{k=1 \\ k\neq j}}^{n} b_{2k} \operatorname{per}(B(1,2|j,k)) \\ &= \frac{(n-2)!}{n^{n-2}} \sum_{\substack{k=1 \\ k\neq j}}^{n} b_{2k} \\ &= \frac{(n-2)!}{n^{n-2}}(1-b_{2j}), \end{aligned}$$

$j = 1, \dots, n$. It follows that $b_{21} = b_{22} = \cdots = b_{2n} = 1/n$, and therefore also $b_{11} = b_{12} = \cdots = b_{1n} = 1/n$. Hence $B = J_n$, contradicting the hypothesis that s is the least positive integer such that $M_n^s A = J_n$. We conclude that all the terms in the sequence (1.17) are different from J_n.

Therefore (1.17) is an infinite sequence of positive minimizing matrices different from J_n. Since $M_n^t \to J_n$, this implies that in any arbitrarily small neighborhood of J_n there are matrices distinct from J_n with the same permanent as that of J_n, contradicting Theorem 1.4. Hence the assumption that $A \neq J_n$ cannot hold. The theorem is proved. ■

ISBN 0-201-13505-1

Sasser and Slater [168] have actually shown that in the interior of Ω_n the only relative minimum for the permanent function occurs at J_n.

The van der Waerden conjecture is generally believed to be true, although there is no strong evidence in support of this conclusion. The conjecture cannot be considered as "intuitively obvious": It is not at all obvious that the absolute minimum for the permanent function in Ω_n cannot be attained on the boundary of Ω_n, nor that such a minimum must be unique. Indeed, neither assertion is true in certain subpolyhedra of Ω_n. For example, let P_1, P_2, P_3, and P_4 be the three-square permutation matrices corresponding to the permutations (123), (132), (12), and (13), and let

$$\Gamma=\left\{A\in\Omega_3 | A=\sum_{j=1}^{4}\theta_j P_j, \qquad \theta_j\geqslant 0, \qquad \sum_{j=1}^{4}\theta_j=1\right\}.$$

In other words, Γ consists of all 3×3 doubly stochastic matrices with 0 in the (1, 1) position. It is easy to show (see Problem 2) that

$$\operatorname{per}(A)\geqslant\tfrac{1}{4}$$

for any A in Γ, and that the minimum is attained for *all* matrices of the form

$$\begin{bmatrix} 0 & \frac{1}{2} & \frac{1}{2} \\ \frac{1}{2} & \theta & \frac{1}{2}-\theta \\ \frac{1}{2} & \frac{1}{2}-\theta & \theta \end{bmatrix}, \tag{1.18}$$

$0\leqslant\theta\leqslant\frac{1}{2}$, including the matrices on the boundary (for $\theta=0$ and $\theta=\frac{1}{2}$). Note that all the subpermanents of order 2 of the matrix (1.18) equal $\frac{1}{4}$, except the permanent of the submatrix in the last two rows and the last two columns, which can take all values between $\frac{1}{8}$ and $\frac{1}{4}$.

5.2 Properties of Minimizing Matrices

We have seen that, if $A\in\Omega_n$ is a positive minimizing matrix, then $A=J_n$ (Theorem 1.5). In this section we study the properties of minimizing matrices different from J_n (if any). It is hoped that this type of study will lead either to a construction of a minimizing matrix other than J_n (which would disprove at least the part of the van der Waerden conjecture that asserts the uniqueness of the minimizing matrix) or to a proof that a minimizing matrix cannot have any zero entries (which would prove the conjecture). At present the results are too fragmentary to allow any hope that the van der Waerden conjecture will be resolved in this manner.

ISBN 0-201-13505-1

Theorem 1.2 and the techniques developed in the preceding section can be utilized in obtaining significant information about the location of zeros (if any) in a minimizing matrix.

THEOREM 2.1 [62]. *If $A \neq J_n$ is a minimizing matrix, then all its zeros cannot occur in a single row (column).*

Proof. We show that the assumption that a minimizing matrix $A = (a_{ij}) \neq J_n$ has all its zeros in the same row (or in the same column) leads to a contradiction. Suppose then that $a_{11} = a_{12} = \cdots = a_{1r} = 0$, and that $a_{ij} > 0$ otherwise. Clearly $r \leqslant n-2$, since A is totally indecomposable. Let $N = 1 \dotplus M_{n-1}$, where M_{n-1} is the matrix (1.12). Then, since $\operatorname{per}(A(i|j)) = \operatorname{per}(A)$, $i = 2, \ldots, n$, $j = 1, \ldots, n$, we can show by a reasoning similar to that in the proof of Theorem 1.3 that

$$\operatorname{per}(NA) = \operatorname{per}(A),$$

so that NA is again a minimizing matrix with the same first row as A. Thus $N^t A$ is a minimizing matrix with first row equal to $A_{(1)}$ for all positive integers t. Since $\lim_{t\to\infty} N^t = 1 \dotplus J_{n-1}$, the same conclusion must hold true for the matrix

$$L = (1 \dotplus J_{n-1})A$$

$$= \begin{bmatrix} 0 & \cdots & 0 & a_{1,r+1} & \cdots & a_{1n} \\ \dfrac{1}{n-1} & \cdots & \dfrac{1}{n-1} & \dfrac{1-a_{1,r+1}}{n-1} & \cdots & \dfrac{1-a_{1n}}{n-1} \\ \vdots & & \vdots & \vdots & & \vdots \\ \dfrac{1}{n-1} & \cdots & \dfrac{1}{n-1} & \dfrac{1-a_{1,r+1}}{n-1} & \cdots & \dfrac{1-a_{1n}}{n-1} \end{bmatrix}.$$

Now, $\operatorname{per}(L(1|j)) = \operatorname{per}(L)$, $j = r+1, \ldots, n$. Thus

$$\frac{(n-1)!}{(n-1)^{n-1}} \prod_{\substack{t=r+1 \\ t\neq j}}^{n} (1 - a_{1t}) = \operatorname{per}(L).$$

It follows that

$$1 - a_{1,r+1} = 1 - a_{1,r+2} = \cdots = 1 - a_{1n};$$

that is,

$$a_{1,r+1} = a_{1,r+2} = \cdots = a_{1n} = \frac{1}{n-r}.$$

ISBN 0-201-13505-1

Thus

$$\begin{aligned}\operatorname{per}(L) &= \frac{(n-1)!}{(n-1)^{n-1}}\left(1-\frac{1}{n-r}\right)^{n-r-1}\\ &= \operatorname{per}(J_n)\left(1-\frac{1}{n-r}\right)^{n-r-1}\Big/\left(1-\frac{1}{n}\right)^{n-1}. \qquad (2.1)\end{aligned}$$

Now, the function $f(s)=(1-1/s)^{s-1}$ is monotonically decreasing, and therefore

$$\frac{f(n-r)}{f(n)}=\left(1-\frac{1}{n-r}\right)^{n-r-1}\Big/\left(1-\frac{1}{n}\right)^{n-1}>1.$$

It follows from (2.1) that

$$\operatorname{per}(L)>\operatorname{per}(J_n),$$

which contradicts the fact that L is a minimizing matrix. We conclude that A cannot have all its zeros in the same row or in the same column. ∎

Example 2.1 [62]. Use Theorems 1.3 and 2.1 to prove the van der Waerden conjecture for $n=3$.

Suppose that A is a minimizing matrix different from J_3. Then A must be totally indecomposable, and it cannot have all its zeros in the same row or in the same column. Thus, there exist permutation matrices P and Q such that

$$PAQ=\begin{bmatrix} 0 & a & 1-a \\ b & 0 & 1-b \\ 1-b & 1-a & a+b-1 \end{bmatrix},$$

where $0<a<1$, $0<b<1$, and $a+b\geqslant 1$. By Theorem 1.3,

$$\operatorname{per}(PAQ\,(2|3))=\operatorname{per}(PAQ\,(3|2))=\operatorname{per}(A).$$

Hence

$$a(1-b)=b(1-a),=\operatorname{per}(A) \qquad (2.2)$$

and therefore $a=b$. Suppose that $a+b-1>0$. Then

$$\operatorname{per}(PAQ\,(3|3))=\operatorname{per}(A);$$

that is,

$$ab=a^2=\operatorname{per}(A). \qquad (2.3)$$

ISBN 0-201-13505-1

From (2.2) and (2.3) we deduce that $a=b=\frac{1}{2}$, which contradicts the assumption that $a+b-1>0$. However, if $a+b-1=0$, then we again obtain $a=b=\frac{1}{2}$. Thus, if A is a minimizing matrix different from J_n, then

$$PAQ=\begin{bmatrix} 0 & \frac{1}{2} & \frac{1}{2} \\ \frac{1}{2} & 0 & \frac{1}{2} \\ \frac{1}{2} & \frac{1}{2} & 0 \end{bmatrix}. \tag{2.4}$$

But then $\text{per}(A)=1/4>\text{per}(J_3)=2/9$, which contradicts the hypothesis that A is a minimizing matrix. Hence J_3 must be the unique minimizing matrix in Ω_3.

Our next theorem, due to London [199], improves a result of Eberlein [152], which in turn was a generalization of Theorem 2.1. We begin with three preliminary results.

LEMMA 1 [199]. *Let $A\in\Omega_n$ be a minimizing matrix. If rows $A_{(1)}$ and $A_{(2)}$ have the same zero pattern, then*

$$\text{per}(A)=\text{per}\left(\tfrac{1}{2}(A_{(1)}+A_{(2)}),\tfrac{1}{2}(A_{(1)}+A_{(2)}),A_{(3)},\ldots,A_{(n)}\right).$$

The proof is analogous to the first part of the proof of Theorem 1.3.

LEMMA 2 [199]. *Let $A\in\Omega_n$ be a minimizing matrix. If the rows $A_{(1)},\ldots,A_{(k)}$ have the same zero pattern, and the remaining rows $A_{(k+1)},\ldots,A_{(n)}$ have the same zero pattern, then*

$$\text{per}(A)=\text{per}(B),$$

where

$$B_{(i)}=\begin{cases} \dfrac{A_{(1)}+\cdots+A_{(k)}}{k} & \text{for } i=1,\ldots,k, \\ \dfrac{A_{(k+1)}+\cdots+A_{(n)}}{n-k} & \text{for } i=k+1,\ldots,n. \end{cases} \tag{2.5}$$

Proof. The lemma can be proved from Lemma 1 by the technique used in proving Theorem 1.5 from Theorem 1.2 (see Problem 4). ■

LEMMA 3 [152]. *Let $B\in\Omega_n$ be a matrix of the form* (2.5); *that is, $B_{(1)}=\cdots=B_{(k)}$ and $B_{(k+1)}=\cdots=B_{(n)}$. Then B cannot be a minimizing matrix unless $B=J_n$.*

Proof. The lemma can be proved much in the same way as Theorem 2.1 (see Problems 5 and 6) or the proof of Theorem 7 in [139]. ■

ISBN 0-201-13505-1

THEOREM 2.2 [199]. *If $A \in \Omega_n$ is a minimizing matrix, $A \neq J_n$, then the rows (columns) of A are of at least three different zero patterns.*

Proof. Suppose that the rows (columns) of A are of two different zero patterns. We can assume without loss of generality that $A_{(1)},\ldots,A_{(k)}$ have the same zero pattern, and so have $A_{(k+1)},\ldots,A_{(n)}$. By Lemma 2, $\operatorname{per}(A)=\operatorname{per}(B)$, where B is the matrix defined in (2.5). But then, by Lemma 3, A cannot be a minimizing matrix. ■

It was shown in [62] that, if A is a minimizing matrix, then $\operatorname{per}(A(i|j))=\operatorname{per}(A)$ if $a_{ij}>0$, and it was asserted that $\operatorname{per}(A(i|j))=\operatorname{per}(A)+\beta$ if $a_{ij}=0$, where β is a nonnegative constant independent of i and j. (See the footnote in [128].) The following result due to London [199] is a somewhat weaker version of the above unproved assertion.

THEOREM 2.3 [199]. *If A is a minimizing matrix in Ω_n, then* $\operatorname{per}(A(i|j))\geqslant \operatorname{per}(A)$ *for all i and j.*

Proof [269]. For any $n\times n$ permutation matrix $P=(p_{ij})$ and $0\leqslant\theta\leqslant 1$, define

$$f_P(\theta)=\operatorname{per}((1-\theta)A+\theta P).$$

Since A is a minimizing matrix,

$$f'_P(0)\geqslant 0$$

for any permutation matrix P. But

$$\begin{aligned} f'_P(0) &= \sum_{s,t=1}^{n}(-a_{st}+p_{st})\operatorname{per}(A(s|t)) \\ &= \sum_{s,t=1}^{n} p_{st}\operatorname{per}(A(s|t))-n\operatorname{per}(A) \\ &= \sum_{s=1}^{n}\operatorname{per}(A(s|\sigma(s)))-n\operatorname{per}(A), \end{aligned}$$

where σ is the permutation corresponding to P. Hence

$$\sum_{s=1}^{n}\operatorname{per}(A(s|\sigma(s)))\geqslant n\operatorname{per}(A) \tag{2.6}$$

for any permutation σ. Now, the matrix A must be fully indecomposable (see Theorem 1.1) and therefore by Theorem 3.5, Chapter 3, any entry of A lies on a diagonal all of whose other entries are positive. In other words, for any (i,j) there exists a permutation σ such that $j=\sigma(i)$ and $a_{s,\sigma(s)}>0$ for

ISBN 0-201-13505-1

$s=1,\ldots,i-1,\ i+1,\ldots,n$. But by Theorem 1.2, this implies that

$$\operatorname{per}\big(A(s|\sigma(s))\big)=\operatorname{per}(A) \tag{2.7}$$

for $s=1,\ldots,i-1,\ i+1,\ldots,n$. It now follows from (2.6) and (2.7) that

$$\operatorname{per}(A(i|j))\geqslant \operatorname{per}(A),$$

since $j=\sigma(i)$. ■

5.3 Some Partial Results. Friedland's Theorem

Marcus and Newman [79] have shown that the van der Waerden conjecture holds for positive semi-definite symmetric doubly stochastic matrices. (See Corollary 1, Section 4.5.) This result was improved by Marcus and Minc as follows.

THEOREM 3.1 [144]. *If A is an n-square positive definite symmetric doubly stochastic matrix with least eigenvalue λ_n, then*

$$\operatorname{per}(A)\geqslant \operatorname{per}(J_n)+\lambda_n^n(1-\operatorname{per}(J_n)).$$

The Marcus–Newman result was extended by Sasser and Slater to a larger class of doubly stochastic matrices.

THEOREM 3.2 [134]. *If A is a normal doubly stochastic matrix with eigenvalues $\lambda_1,\ldots,\lambda_n$ that satisfy*

$$-\pi/2n\leqslant \operatorname{amp}\lambda_i\leqslant \pi/2n,$$

$i=1,\ldots,n$, and $A\neq J_n$, then

$$\operatorname{per}(A)>\operatorname{per}(J_n).$$

This result was also improved by Marcus and Minc.

THEOREM 3.3 [144]. *If A satisfies the hypotheses of Theorem 3.2, then*

$$\operatorname{per}(A)\geqslant \operatorname{per}(J_n)+\frac{1}{2}\,\frac{(n-2)!}{n^{n-2}}\,\|A-J_n\|^2,$$

where $\|A-J_n\|$ denotes the Euclidean norm of $A-J_n$.

Theorem 3.2 was extended by Friedland [243] to the set of all doubly stochastic $n\times n$ matrices, not necessarily normal, whose numerical range lies in the closed sector $[-\pi/2n,\pi/2n]$.

ISBN 0-201-13505-1

We omit the proofs of Theorems 3.1, 3.2, and 3.3. The proofs of Theorems 3.1 and 3.3 require rather involved techniques of multilinear algebra, while the Marcus–Newman and Sasser–Slater theorems follow directly from Friedland's result. In the remaining part of this section we shall give a detailed account of Friedland's paper [243].

DEFINITION 3.1. If $A \in M_{m,n}(\mathbf{C})$ and $1 \leqslant r \leqslant \min(m,n)$, then the rth *induced matrix* of A, denoted by $P_r(A)$, is the $\binom{m+r-1}{r} \times \binom{n+r-1}{r}$ matrix whose entries are

$$\operatorname{per}(A[\alpha|\beta])/\sqrt{\mu(\alpha)\mu(\beta)}$$

arranged lexicographically in $\alpha \in G_{r,m}$ and $\beta \in G_{r,n}$.

If $A \in M_{m,n}(\mathbf{C})$ and $B \in M_{n,p}(\mathbf{C})$, then

$$P_r(AB) = P_r(A)P_r(B) \tag{3.1}$$

(see [92]).

Let $u = (u_1, \ldots, u_n)$ be a nonzero complex n-tuple. We define the $n \times n$ matrix H_u as follows:

$$H_u u = u,$$

and

$$H_u v = 0 \qquad \text{for all } v \in \langle u \rangle^{\perp}.$$

Clearly H_u is a positive semi-definite hermitian matrix with eigenvalues $1, 0, \ldots, 0$. The converse is also true: If A is a hermitian matrix with eigenvalues $1, 0, \ldots, 0$, then $A = H_u$, where u is uniquely determined up to a scalar multiple (see Problem 7).

Let $M_n(u)$ be the subset of $M_n(\mathbf{C})$ defined by

$$M_n(u) = \{A \in M_n(\mathbf{C}) | Au = A^*u = u\}. \tag{3.2}$$

LEMMA 1. *Let $A \in M_n(u)$ and let $B = (A + A^*)/2$. If B is positive semi-definite, then so is $B - H_u$.*

Proof. Clearly $B \in M_n(u)$. Also,

$$(B - H_u)u = 0$$

and

$$((B - H_u)v, v) = (Bv, v)$$
$$\geqslant 0$$

for any $v \in \langle u \rangle^{\perp}$. Hence $B - H_u$ is positive semi-definite. ∎

ISBN 0-201-13505-1

If $u_1,\ldots,u_n$ are n-tuples, the symmetric product $u_1 \quad \cdots \quad u_n$ (see Section 2.2) can be defined as follows. Let G be the $r\times n$ matrix whose ith row is u_i, $i=1,\ldots,r$. The symmetric product $u_1 \quad \cdots \quad u_r$ is the $\binom{n+r-1}{r}$-tuple whose α coordinate is

$$\operatorname{per}(G[1,\ldots,r|\alpha])/\sqrt{\mu(\alpha)}$$

arranged lexicographically in $\alpha\in G_{r,n}$. If C is an $m\times n$ matrix, then

$$P_r(C)u_1*\cdots*u_r=(Cu_1)*\cdots*(Cu_r)$$

(see [92]).

LEMMA 2. *If* $A\in M_n(u)$ *and* $P_r(A)+P_r(A^*)$ *is positive semi-definite, then so is*

$$\frac{P_r(A)+P_r(A^*)}{2}-P_r(H_u).$$

Proof. The matrix $P_r(A)$ belongs to $M_k(u*\cdots*u)$, where $k=\binom{n+r-1}{r}$. For,

$$\begin{aligned}P_r(A)u*\cdots*u&=(Au)*\cdots*(Au)\\&=u*\cdots*u,\end{aligned}$$

and

$$\begin{aligned}P_r(A^*)u*\cdots*u&=(A^*u)*\cdots*(A^*u)\\&=u*\cdots*u.\end{aligned}$$

Also, $P_r(H_u)$ has exactly one nonzero eigenvalue 1, and

$$\begin{aligned}P_r(H_u)u*\cdots*u&=(H_uu)*\cdots*(H_uu)\\&=u*\cdots*u.\end{aligned}$$

Thus $P_r(H_u)=H_{u*\cdots*u}$, and the result follows by Lemma 1. ■

Let $e=(1,\ldots,1)$. Then clearly $H_e=J_n$ and $M_n(e)$ is the set of all $n\times n$ complex matrices all of whose row sums and column sums are equal to 1. In particular, $M_n(e)$ contains all doubly stochastic matrices. Lemma 2 implies the following result.

THEOREM 3.4. *Let* A *be an* $n\times n$ *matrix all of whose row sums and column sums are equal to* 1, *and let* α *be a sequence in* $G_{r,n}$. *If* $P_r(A)+P_r(A^*)$ *is*

ISBN 0-201-1350-1

positive semi-definite hermitian, then

$$\operatorname{Re}\big(\operatorname{per}(A[\alpha|\alpha])\big) \geqslant r!/n^r. \tag{3.3}$$

Proof. Since, by Lemma 2, the matrix $P_r(A)+P_r(A^*)-2P_r(J_n)$ is positive semi-definite, its main diagonal entries are nonnegative. Hence

$$\operatorname{per}(A[\alpha|\alpha])+\operatorname{per}\big(\bar{A}[\alpha|\alpha]\big)-2(r!/n^r)\geqslant 0. \qquad \blacksquare$$

COROLLARY 1. *If A is a positive semi-definite symmetric doubly stochastic $n\times n$ matrix and $\alpha\in G_{r,n}$, then*

$$\operatorname{per}(A[\alpha|\alpha])\geqslant r!/n^r$$

(cf. [124]).

We shall now introduce a larger class of matrices for which the hypothesis of Theorem 3.4 is satisfied.

For a positive number r let X_r and X_r° be the sets of $n\times n$ complex matrices whose numerical ranges lie in the sectors $[-\pi/2r,\pi/2r]$ and $(-\pi/2r,\pi/2r)$, respectively; that is,

$$X_r=\{A\in M_n(\mathbf{C})|\operatorname{Re}(Ax,x)\geqslant|\operatorname{Im}(Ax,x)|\cot\pi/2r\},$$

and

$$X_r^\circ=\{A\in M_n(\mathbf{C})|\operatorname{Re}(Ax,x)>|\operatorname{Im}(Ax,x)|\cot\pi/2r\}.$$

It is easy to see that X_r and X_r° can be also described as follows: X_r is the set of all complex $n\times n$ matrices A such that $A+A^*\pm i(A-A^*)\cot\pi/2r$ is positive semi-definite; X_r° is the subset of X_r consisting of matrices A such that $A+A^*\pm i(A-A^*)\cot\pi/2r$ is positive definite.

THEOREM 3.5. *Let $A\in M_n(\mathbf{C})$. Then $P_r(A)+P_r(A^*)$ is positive definite if and only if $\omega A\in X_r^\circ$ for some* rth *root of unity ω.*

Proof. For $K\in M_r(\mathbf{C})$, let $D(K)$ denote the numerical range of K; that is,

$$D(K)=\{z\in\mathbf{C}|z=(Kx,x),\|x\|=1\}$$

and let

$$D^r(K)=\{z|z=\zeta^r,\zeta\in D(K)\}.$$

Note that $D(K)$ is a compact connected domain. Also $D^r(K)\subset D(P_r(K))$,

ISBN 0-201-1350-1

since

$$(P_r(K)x*\cdots *x,x*\cdots *x)=(Kx,x)^r.$$

Now assume that $P_r(A)+P_r(A^*)$ is positive definite. This implies that $D^r(A)$ is contained in the open right half plane, and since $D(A)$ is connected it must be contained in the sector

$$\{z|-\pi/2r<\mathrm{amp}(z\omega)<\pi/2r\}$$

for some rth root of unity ω. Thus, for any $x\neq 0$,

$$\mathrm{Re}(\omega Ax,x)>|\mathrm{Im}(\omega Ax,x)|\cot(\pi/2r);$$

that is, $\omega A\in X_r^\circ$.

To prove the sufficiency let

$$B=\omega A+\bar{\omega}A^* \qquad \text{and} \qquad C=i(\omega A-\bar{\omega}A^*).$$

Since $\omega A\in X_r^\circ$, the matrix B is positive definite hermitian. Hence (see Problem 9) there exists a nonsingular matrix Y such that

$$B=Y^*Y \qquad \text{and} \qquad C=Y^*DY,$$

where $D=\mathrm{diag}(d_1,\ldots,d_n)$. Since $B\pm\cot(\pi/2r)C$ is positive definite, we have

$$\tan(\pi/2r)>|d_j|$$

for $j=1,\ldots,n$. Let $G=\mathrm{diag}(g_1,\ldots,g_n)=\frac{1}{2}(I_n-iD)$. Then

$$\omega A=Y^*GY$$

and

$$|\mathrm{amp}(g_j)|<\pi/2r,$$

$j=1,\ldots,n$. Hence, if $\alpha=(\alpha_1,\ldots,\alpha_r)\in G_{r,n}$,

$$\left|\mathrm{amp}\prod_{t=1}^{r} g_{\alpha_t}\right|<\frac{\pi}{2}.$$

It follows that $P_r(G)+P_r(G^*)$ is positive definite. But then

$$P_r(\omega A)+P_r(\bar{\omega}A^*)=P_r(Y^*)(P_r(G)+P_r(G^*))P_r(Y)$$

ISBN 0-201-1350-1

is positive definite, and therefore $P_r(\omega A)+P_r(\bar{\omega}A^*)=P_r(A)+P_r(A^*)$ is positive definite. ■

COROLLARY 2. *If A belongs to the set X_r, then $P_r(A)+P_r(A^*)$ is positive semi-definite hermitian.*

The corollary follows from Theorem 3.5 by a continuity argument. An immediate consequence of Theorem 3.4 and Corollary 2 is the following result.

THEOREM 3.6. *If A is an $n\times n$ doubly stochastic matrix whose numerical range is contained in the closed sector $[-\pi/2r,\pi/2r]$ and if $\alpha\in G_{r,n}$, then*

$$\operatorname{per}(A[\alpha|\alpha])\geqslant r!/n^r.$$

In particular, if $r=n$ and $\alpha=(1,\dots,n)$ then

$$\operatorname{per}(A)\geqslant n!/n^n.$$

We omit the discussion of the case of equality.

5.4 A Conjecture of Marcus and Minc

In the present section we propose a conjecture of Marcus and Minc [128] and prove it for two classes of doubly stochastic matrices for which the van der Waerden conjecture is known to hold.

CONJECTURE. *If S is a doubly stochastic $n\times n$ matrix, $n\geqslant 2$, then*

$$\operatorname{per}(S)\geqslant\operatorname{per}\left(\frac{nJ_n-S}{n-1}\right). \tag{4.1}$$

If $n\geqslant 4$, equality can hold in (4.1) *if and only if $S=J_n$.*

For $n=3$ the condition for uniqueness does not hold. (See Problem 19.)

Note that the above conjecture implies the van der Waerden conjecture. For, let $f:\Omega_n\to\Omega_n$ be defined by

$$f(S)=J_n-\frac{S-J_n}{n-1}.$$

Then, if $S\neq J_n$, the inequality (4.1) becomes

$$\operatorname{per}(S)>\operatorname{per}(f(S)),$$

and therefore

$$\operatorname{per}(S)>\operatorname{per}(f(S))>\operatorname{per}(f^k(S))$$

ISBN 0-201-1350-1

for $k=2,3,\ldots$. Hence

$$\begin{aligned}\operatorname{per}(S) &> \lim_{k\to\infty} \operatorname{per}\left(f^k(S)\right)\\ &= \operatorname{per}\left[\lim_{k\to\infty}\left(J_n+(-1)^k\frac{S-J_n}{(n-1)^k}\right)\right]\\ &= \operatorname{per}(J_n)\\ &= n!/n^n.\end{aligned}$$

We now show that the conjecture holds for doubly stochastic matrices in a sufficiently small neighborhood of J_n and for all positive semi-definite symmetric doubly stochastic n-square matrices (compare Theorem 1.4 and Corollary 1, Section 4.5). First, however, we prove a preliminary result [128].

THEOREM 4.1. *Let $A=(a_{ij})$ be an $n\times n$ real matrix all of whose row sums and column sums are equal to* 0. *Then the sum of all subpermanents of A of order* 2 *is positive unless $A=0$.*

Proof. Let σ_k denote the sum of all subpermanents of A of order k. Then

$$\begin{aligned}\sigma_2 &= \sum_{(i,j)\in Q_{2,n}}\ \sum_{(s,t)\in Q_{2,n}} (a_{is}a_{jt}+a_{js}a_{it})\\ &= \sum_{(i,j)\in Q_{2,n}}\ \sum_{s=1}^{n} a_{is}\sum_{\substack{t=1\\ t\neq s}}^{n} a_{jt}\\ &= -\sum_{(i,j)\in Q_{2,n}}\ \sum_{s=1}^{n} a_{is}a_{js},\end{aligned}$$

since $\sum_{t=1}^{n} a_{jt}=0,\ j=1,\ldots,n$. Thus

$$\begin{aligned}\sigma_2 &= -\sum_{s=1}^{n}\ \sum_{(i,j)\in Q_{2,n}} a_{is}a_{js}\\ &= -\frac{1}{2}\sum_{s=1}^{n}\sum_{i=1}^{n}\sum_{\substack{j=1\\ j\neq i}}^{n} a_{is}a_{js}\\ &= -\frac{1}{2}\sum_{s=1}^{n}\sum_{i=1}^{n} a_{is}(-a_{is})\\ &= \frac{1}{2}\sum_{i,s=1}^{n} a_{is}^2\\ &\geqslant 0,\end{aligned}$$

ISBN 0-201-13505-1

and equality can hold if and only if all a_{is} are zero—that is, if and only if $A=0$. ■

The following two theorems were obtained by Marcus and Minc [128].

THEOREM 4.2. *If $S=(s_{ij})$ is a doubly stochastic matrix in a sufficiently small neighborhood of J_n, then*

$$\operatorname{per}(S)\geqslant\operatorname{per}\left(\frac{nJ_n-S}{n-1}\right). \tag{4.2}$$

If $n=2$, then (4.2) is an equality. If $n\geqslant 3$, then equality holds in (4.2) if and only if $S=J_n$.

Proof. Let $A=(a_{ij})=S-J_n$. Then A satisfies the hypotheses of Theorem 4.1. Hence $\sigma_2>0$ unless $A=0$; that is, $S=J_n$. Now, $\sigma_1=0$, and therefore

$$\begin{aligned}\operatorname{per}(S)&=\operatorname{per}(J_n+A)\\&=\frac{n!}{n^n}+\frac{(n-1)!}{n^{n-1}}\sigma_1+\frac{(n-2)!}{n^{n-2}}\sigma_2+\cdots+\frac{2!}{n^2}\sigma_{n-2}+\frac{1!}{n}\sigma_{n-1}+\sigma_n\\&=\frac{n!}{n^n}+\frac{(n-2)!}{n^{n-2}}\sigma_2+\sum_{t=3}^{n}\frac{(n-t)!}{n^{n-t}}\sigma_t.\end{aligned}$$

On the other hand,

$$\begin{aligned}\operatorname{per}\left(\frac{nJ_n-S}{n-1}\right)&=\operatorname{per}\left(J_n-\frac{A}{n-1}\right)\\&=\frac{n!}{n^n}+\frac{(n-2)!}{n^{n-2}}\frac{1}{(n-1)^2}\sigma_2+\sum_{t=3}^{n}(-1)^t\frac{(n-t)!}{n^{n-t}}\frac{1}{(n-1)^t}\sigma_t.\end{aligned}$$

Hence if $n\geqslant 3$, $A\neq 0$, and if all the entries of A are sufficiently small in absolute value so that

$$\frac{(n-2)!}{n^{n-2}}\left(1-\frac{1}{(n-1)^2}\right)\sigma_2+\sum_{t=3}^{n}\left[1-\frac{(-1)^t}{(n-1)^t}\right]\frac{(n-t)!}{n^{n-t}}\sigma_t>0, \tag{4.3}$$

then

$$\operatorname{per}(S)-\operatorname{per}\left(\frac{nJ_n-S}{n-1}\right)>0.$$

If $S\neq J_n$ is a doubly stochastic matrix in a sufficiently small neighborhood of J_n so that (4.3) holds, then (4.2) cannot be an equality. Of course, if

ISBN 0-201-13505-1

$S=J_n$, then actually

$$S=\frac{nJ_n-S}{n-1}$$

and equality trivially holds in (4.2). ■

THEOREM 4.3. *If S is a positive semi-definite symmetric doubly stochastic matrix, then*

$$\operatorname{per}(S)\geqslant \operatorname{per}\left(\frac{nJ_n-S}{n-1}\right). \tag{4.4}$$

If $n\geqslant 3$, the equality can hold in (4.4) *if and only if $S=J_n$.*

Proof. Let $1=\lambda_1\geqslant\lambda_2\geqslant\cdots\geqslant\lambda_n\geqslant 0$ be the eigenvalues of S. Since S and J_n commute, the eigenvalues of $(nJ_n-S)/(n-1)$ are

$$1,-\frac{\lambda_2}{n-1},\ldots,-\frac{\lambda_n}{n-1}.$$

Let $v_1=(1,\ldots,1)/\sqrt{n}\,,v_2,\ldots,v_n$ be an orthonormal set of eigenvectors common to S and $(nJ_n-S)/(n-1)$, and let U be the unitary matrix whose ith row is v_i, $i=1,\ldots,n$. By formula (5.8), Section 4.5,

$$\operatorname{per}(S)=\sum_{\gamma\in G_{n,n}}\frac{c_\gamma}{\mu(\gamma)}\cdot\prod_{t=2}^{n}\lambda_t^{m_t(\gamma)},$$

and

$$\operatorname{per}\left(\frac{nJ_n-S}{n-1}\right)=\sum_{\gamma\in G_{n,n}}\frac{c_\gamma}{\mu(\gamma)}\prod_{t=2}^{n}\left(-\frac{\lambda_t}{n-1}\right)^{m_t(\gamma)},$$

where $c_\gamma=|\operatorname{per}(U[\gamma|1,\ldots,n])|^2$, $\mu(\gamma)=\prod_{t=1}^{n}m_t(\gamma)!$, and $m_t(\gamma)$ denotes the number of times the integer t occurs in γ. Clearly

$$\frac{c_\gamma}{\mu(\gamma)}\prod_{t=2}^{n}\lambda_t^{m_t(\gamma)}\geqslant\frac{c_\gamma}{\mu(\gamma)}\prod_{t=2}^{n}\left(-\frac{\lambda_t}{n-1}\right)^{m_t(\gamma)} \tag{4.5}$$

for any γ, and hence the inequality (4.4) follows.

If equality holds in (4.4), then (4.5) is equality for every γ. We show by an appropriate choice of γ that this implies that $\lambda_2=\cdots=\lambda_n=0$. Let $v_2=(x_1,\ldots,x_n)$, and suppose that $x_{i_1},\ldots,x_{i_k}$ are nonzero, and $x_j=0$ for

ISBN 0-201-13505-1

$j \notin \{i_1,\ldots,i_k\}$. Let $\gamma=(\gamma_1,\ldots,\gamma_n)$ where $\gamma_1=\cdots=\gamma_{n-k}=1$ and $\gamma_{n-k+1}=\cdots=\gamma_n=2$. Then

$$c_\gamma=\left|\operatorname{per}(U[1,\ldots,1,2,\ldots,2|1,\ldots,n])\right|^2$$

$$=\frac{1}{n^{n-k}}\left|\operatorname{per}\begin{bmatrix}1 & \cdots & 1\\ \vdots & & \vdots\\ 1 & \cdots & 1\\ x_1 & \cdots & x_n\\ \vdots & & \vdots\\ x_1 & \cdots & x_n\end{bmatrix}\right|^2. \tag{4.6}$$

We evaluate the permanent in (4.6) using the Laplace expansion on the last k rows:

$$c_\gamma=\frac{1}{n^{n-k}}\left|k!(n-k)!\left(\prod_{s=1}^{k} x_{i_s}\right)\right|^2.$$

Hence $c_\gamma \neq 0$, and therefore equality in (4.5) implies that

$$\lambda_2^k=\left(-\frac{\lambda_2}{n-1}\right)^k,$$

and thus $\lambda_2=0$. But then $\lambda_2=\cdots=\lambda_n=0$, the doubly stochastic matrix S has rank 1, and therefore $S=J_n$. The converse is obvious. ■

Wang [287] proved inequality (4.1) for all doubly stochastic 3×3 matrices, and conjectured that

$$\operatorname{per}(S)\geqslant \operatorname{per}\left(\frac{nJ_n+S}{n+1}\right)$$

for all $S\in\Omega_n$.

5.5 Lower Bounds for the Permanents of Doubly Stochastic Matrices

Marcus and Newman [62] showed that the van der Waerden conjecture holds for $n=3$ (see Example 2.1). Eberlein and Mudholkar [139] used a different method to prove that it holds for $n=3$ and 4. Eberlein [152] then established the validity of the conjecture for $n=5$. Gleason [177] re-proved

ISBN 0-201-13505-1

cases $n=3$ and $n=4$ using an entirely new approach which, he believes, may settle with the aid of a computer the van der Waerden conjecture for $n=5$ and perhaps larger matrices. There the matter rests at present. For a general n, the bounds that have been proved are considerably lower than $n!/n^n$.

Every doubly stochastic $n\times n$ matrix A can be expressed (Theorem 3.3, Chapter 3) in the form

$$A=\sum_{j=1}^{s}\theta_j P_j,$$

where $P_1,\ldots,P_s$ are permutation matrices and $\theta_1,\ldots,\theta_s$ are positive numbers, $\sum_{j=1}^{s}\theta_j=1$. Hence

$$\begin{aligned}\operatorname{per}(A)&\geqslant \max_j \theta_j^n\\ &\geqslant 1/s^n.\end{aligned} \tag{5.1}$$

Actually, it is easy to see that, if $s>1$, then the inequality in (5.1) is strict. (See Problem 16.) Now, the dimension of the polyhedron Ω_n is $(n-1)^2$ (see Problem 17), and thus every doubly stochastic matrix can be represented as a convex combination of no more than $(n-1)^2+1$ permutation matrices. Hence, by (5.1),

$$\operatorname{per}(A)>1/\big((n-1)^2+1\big)^n \tag{5.2}$$

for any A in Ω_n, $n\geqslant 2$.

Paul Erdös observed that, if the van der Waerden conjecture is true, then for any $A\in\Omega_n$,

$$\begin{aligned}\operatorname{per}(A)&=\sum_{\sigma\in S_n}\prod_{i=1}^{n}a_{i,\sigma i}\\ &\geqslant n!/n^n,\end{aligned}$$

and therefore,

$$\max_\sigma \prod_{i=1}^{n}a_{i,\sigma i}\geqslant\frac{1}{n^n}.$$

In other words, the van der Waerden conjecture implies that every doubly stochastic $n\times n$ matrix contains a diagonal the product of whose elements is at least $1/n^n$. This fact is proved in the next theorem.

ISBN 0-201-13505-1

THEOREM 5.1 [77]. *For any doubly stochastic $n\times n$ matrix $S=(s_{ij})$ there exists a permutation σ such that*

$$\prod_{i=1}^{n} s_{i,\sigma i} \geqslant \frac{1}{n^n}. \tag{5.3}$$

Equality can hold in (5.3) *if and only if $S=J_n$.*

Proof. Define

$$f(t)=t\log t$$

if $t>0$, and $f(0)=0$. Then f is strictly convex on the closed interval $[0,1]$. Hence the function F on Ω_n defined by

$$F(S)=\sum_{i,j=1}^{n} f(s_{ij})$$

is strictly convex on Ω_n; that is,

$$F(\theta S+(1-\theta)T)<\theta F(S)+(1-\theta)F(T),$$

$0<\theta<1$, unless $S=T$. Thus, if P is the permutation matrix with 1's in the positions $(1,2),\ldots,(n-1,n),(n,1)$, and $S\in\Omega_n$, then, since $F(P^tS)=F(S)$, we have

$$\begin{aligned} F(S)&=\frac{1}{n}\sum_{t=0}^{n-1} F(P^tS)\\ &\geqslant F\left(\sum_{t=0}^{n-1}\frac{P^tS}{n}\right)\\ &=F(J_nS)\\ &=F(J_n)\\ &=n\log(1/n), \end{aligned}$$

with equality if and only if $P^tS=S$ for all t—that is, if and only if $S=J_n$.

Now by the corollary to Theorem 3.2, Chapter 3, the matrix S has a positive diagonal. Let X be the set of permutations σ for which the diagonals $(a_{1,\sigma 1},\ldots,a_{n,\sigma n})$ are positive, and let Y be the set of corresponding

ISBN 0-201-13505-1

permutation matrices. Denote the convex hull of Y by H. Then $S\in H$ and

$$\begin{aligned}\max_{\sigma\in X}\sum_{i=1}^{n}\log s_{i,\sigma i}&=\max_{P\in Y}\sum_{i,j=1}^{n}p_{ij}\log s_{ij}\\&=\max_{T\in H}\sum_{i,j=1}^{n}t_{ij}\log s_{ij}\\&\geqslant\sum_{i,j=1}^{n}f(s_{ij})\\&=F(S)\\&\geqslant n\log(1/n).\end{aligned}$$

It follows that

$$\max_{\sigma\in S_n}\prod_{i=1}^{n}s_{i,\sigma i}\geqslant\frac{1}{n^n},\tag{5.4}$$

with equality if and only if $S=J_n$. For, if equality holds in (5.4), then $F(S)=n\log(1/n)$, and thus $S=J_n$. The converse is obvious. ■

Theorem 5.1 can be improved as follows [101].

THEOREM 5.2. *Let $S=(s_{ij})$ be a doubly stochastic $n\times n$ matrix with h eigenvalues of modulus* 1. *Then*

$$\max_{\sigma\in S_n}\prod_{i=1}^{n}s_{i,\sigma i}\geqslant\frac{1}{(n-h+1)^{n-h+1}},\tag{5.5}$$

where equality holds if and only if $PSQ=I_{h-1}\dot{+}J_{n-h+1}$ for some permutation matrices P and Q. If S happens to be irreducible, then

$$\max_{\sigma\in S_n}\prod_{i=1}^{n}s_{i,\sigma i}\geqslant\left(\frac{h}{n}\right)^n.\tag{5.6}$$

We omit the proof.

Of course, the inequalities (5.3), (5.5), and (5.6) provide lower bounds for per(S). In fact, these were the best known bounds, for general n, until 1972, when Rothaus [216] announced the following improved bound.

THEOREM 5.3. *If S is a doubly stochastic $n\times n$ matrix, then*

$$\operatorname{per}(S)\geqslant 1/n^{n-1}.\tag{5.7}$$

ISBN 0-201-13505-1

Friedland suggested a simple proof [unpublished] of Theorem 5.3 by applying the inequality (5.3) to a minimizing matrix. Let

$$\operatorname{per}(A)=\min_{S\in\Omega_n}\operatorname{per}(S).$$

We can assume without loss of generality that, by Theorem 5.1,

$$\prod_{t=1}^{n} a_{tt}\geqslant\frac{1}{n^n}.$$

Now, by Theorem 1.2,

$$\operatorname{per}(A(i|i))=\operatorname{per}(A),$$

$i=1,\ldots,n$. Clearly

$$\begin{aligned}\operatorname{per}(A(i|i))&\geqslant\prod_{t\neq i} a_{tt}\\&=a_{ii}^{-1}\prod_{t=1}^{n} a_{tt}.\end{aligned}$$

Hence

$$\begin{aligned}(\operatorname{per}(A))^n&=\prod_{i=1}^{n}\operatorname{per}(A(i|i))\\&\geqslant\prod_{i=1}^{n}\left(a_{ii}^{-1}\prod_{t=1}^{n}a_{tt}\right)\\&=\left(\prod_{t=1}^{n}a_{tt}\right)^{n-1}\\&\geqslant 1/n^{n(n-1)}.\end{aligned}$$

Recently Friedland [296] obtained a substantially sharper bound:

$$\operatorname{per}(S)\geqslant 1/n!, \tag{5.8}$$

for all $S\in\Omega_n$. In 1976 Bang [274] announced the inequality:

$$\operatorname{per}(S)\geqslant 1/e^{n-1} \tag{5.9}$$

for all $S\in\Omega_n$. This bound is virtually of the same order of magnitude as the conjectured bound of van der Waerden. At the time of writing no proof of (5.9) has been published.

ISBN 0-201-13505-1

Problems

1. Let A be a *locally minimizing matrix*—that is, $\operatorname{per}(A) \leqslant \operatorname{per}(S)$ for all doubly stochastic $n \times n$ matrices S in some neighborhood of A. Show that Theorems 1.1, 1.2, and 2.3 hold for locally minimizing matrices.
2. Let Γ be the subpolyhedron of Ω_4 described at the conclusion of Section 5.1. Show that

$$\operatorname{per}(A) \geqslant \tfrac{1}{4}$$

for any A in Γ.
3. Let Δ be the subpolyhedron of Ω_n consisting of matrices $A = (a_{ij})$ with $a_{13} = a_{14} = a_{22} = a_{24} = a_{31} = a_{41} = 0$. Show that

$$\operatorname{per}(A) \geqslant \tfrac{1}{8} \tag{*}$$

for any A in Δ, and find all the matrices in Δ for which (*) is equality.
4. Prove in detail Lemma 2, Section 5.2.
5. Let B be an $n \times n$ doubly stochastic minimizing matrix, and suppose that the first k rows of B are equal and the last $n-k$ rows are equal. Show without the use of Lemma 3, Section 5.2, that either the nonzero entries in the first k rows are equal or the nonzero entries in the last $n-k$ rows are equal. (*Hint*. If $B \neq J_n$, we can assume without loss of generality that $B_{(i)} = (x_1, \ldots, x_s, 0, \ldots, 0)$, $i = 1, \ldots, k$, where $x_j > 0$, $j = 1, \ldots, s$. Use the result in Theorem 1.2 to prove that $x_1 = \cdots = x_s$.)
6. Let $B \in \Omega_n$ and $B_{(1)} = \cdots = B_{(k)} = (\frac{1}{s}, \ldots, \frac{1}{s}, 0, \ldots, 0)$, $B_{(k+1)} = \cdots = B_{(n)}$. Show that $\operatorname{per}(B) > \operatorname{per}(J_n)$.
7. Let A be a hermitian matrix with eigenvalues $1, 0, \ldots, 0$. Show that $A = H_u$, where H_u is the matrix defined in Section 5.3, for some vector u uniquely determined to within a scalar multiple.
8. Prove that, if H is positive definite hermitian, then so is $P_r(H)$.
9. Let H and K be hermitian matrices. Show that, if H is positive definite, then H and K are simultaneously congruent to I_n and a diagonal matrix, respectively.
10. Let

$$X_r = \{A \in M_n(\mathbf{C}) \mid \operatorname{Re}(Ax, x) \geqslant |\operatorname{Im}(Ax, x)| \cot \pi/2r\}$$

(see Section 5.3). Show that X_r is the set of all complex $n \times n$ matrices A such that $A + A^* \pm i(A - A^*) \cot \pi/2r$ is positive semi-definite.
11. Prove Corollary 2, Section 5.3, in detail.
12. Show that Theorem 3.6 implies Theorem 3.2.
13. Investigate the Marcus–Minc conjecture in Section 5.4 for the case $n = 2$.
14. Prove the validity of the Marcus–Minc conjecture for doubly

ISBN 0-201-13505-1

stochastic matrices of the form

$$\begin{bmatrix} 0 & a & 1-a \\ 1-a & 0 & a \\ a & 1-a & 0 \end{bmatrix}.$$

15. Prove the validity of the Marcus–Minc conjecture for doubly stochastic matrices of the form

$$\begin{bmatrix} 0 & a & 1-a \\ b & 0 & 1-b \\ 1-b & 1-a & a+b-1 \end{bmatrix}.$$

16. Prove that, if $A=\Sigma_{j=1}^{s}\theta_j P_j$, where the P_j are permutation matrices and the θ_j are positive numbers such that $\Sigma_{j=1}^{s}\theta_j=1$, then $\operatorname{per}(A)\geqslant 1/s^{n-1}$.
17. Prove that the dimension of Ω_n is $(n-1)^2$.
18. Show that the following conjecture implies the van der Waerden conjecture.
    CONJECTURE. *If* $A\in\Omega_n$ *is a minimizing matrix, then* $\operatorname{per}(A(i|j))=\operatorname{per}(A)$ *for all* i,j.
19. Find matrix $S\in\Omega_3$ different from J_3 such that

$$\operatorname{per}(S)=\operatorname{per}\left(\frac{3J_3-S}{2}\right).$$

ISBN 0-201-13505-1

CHAPTER 6

Upper Bounds for Permanents

6.1 From Muir to Jurkat and Ryser

Muir [14] observed that, if $A=(a_{ij})$ is a 3×3 matrix, then

$$\operatorname{per}(A)=\prod_{s=1}^{3}(a_{s1}i_1+a_{s2}i_2+a_{s3}i_3),$$

where $i_1i_2i_3=1$, and i_1,i_2,i_3 are "symbols subject to the laws of ordinary algebra, except that $i_1^2=i_2^2=i_3^2=0$." He further commented that "it is almost the same to say that per(A) is the coefficient of xyz in the expansion of the product

$$\prod_{s=1}^{3}(a_{s1}x+a_{s2}y+a_{s3}z)."$$

There can be little doubt that Muir used the 3×3 matrix merely as an illustration and that he was aware that his formula can be extended to any square matrix. In fact, if $i_1,\ldots,i_n$ is a basis of an associative linear algebra over a field F such that

$$i_1i_2\cdots i_n\neq 0, \qquad i_si_t=i_ti_s, \qquad \text{and} \quad i_s^2=0,$$

for all s and t, and if $A=(a_{st})$ is an $n\times n$ matrix with entries from F, then

$$\operatorname{per}(A)i_1i_2\cdots i_n=\prod_{s=1}^{n}\sum_{t=1}^{n}a_{st}i_t. \tag{1.1}$$

ENCYCLOPEDIA OF MATHEMATICS and Its Applications, Gian-Carlo Rota (ed.). Vol. 6: Henryk Minc, Permanents

ISBN 0-201-13505-1

Jurkat and Ryser [112] suggested the following realization of equation (1.1) in terms of $2^n \times 2^n$ matrices. Let

$$I=\begin{bmatrix} 1 & 0 \\ 0 & 1 \end{bmatrix} \quad \text{and} \quad E=\begin{bmatrix} 0 & 0 \\ 1 & 0 \end{bmatrix}$$

and let

$$E_i = I \otimes \cdots \otimes I \otimes E \otimes I \otimes \cdots \otimes I,$$

a direct product of n 2×2 matrices with E in the ith position. Clearly, $E_iE_j = E_jE_i$ and $E_i^2=0$, for all i, j, and $E_1E_2\cdots E_n = E_{2^n,1}$ is the $2^n\times 2^n$ matrix with 1 in the $(2^n,1)$ position and 0's elsewhere. Thus, if A is any $n\times n$ matrix, then

$$\operatorname{per}(A)E_1\cdots E_n = \prod_{i=1}^{n} \sum_{j=1}^{n} a_{ij}E_j. \tag{1.2}$$

Define

$$X(A_{(i)}) = X(a_{i1},\ldots,a_{in}) = \sum_{j=1}^{n} a_{ij}E_j. \tag{1.3}$$

Then

$$\begin{aligned} X(a_{i1},\ldots,a_{in}) &= a_{i1}E_1 + \sum_{j=2}^{n} a_{ij}E_j \\ &= a_{i1}E\otimes I\otimes\cdots\otimes I + I\otimes X(a_{i2},\ldots,a_{in}), \end{aligned}$$

where $X(a_{i2},\ldots,a_{in})$ is a $2^{n-1}\times 2^{n-1}$ matrix defined as in (1.3). Hence we have

$$X(a_{i1}) = \begin{bmatrix} 0 & 0 \\ a_{i1} & 0 \end{bmatrix},$$

and for $n \geqslant 2$,

$$X(a_{i1},\ldots,a_{in}) = \left[\begin{array}{c|c} X(a_{i2},\ldots,a_{in}) & 0 \\ \hline a_{i1}I & X(a_{i2},\ldots,a_{in}) \end{array}\right],$$

where I and 0 denote the identity matrix and zero matrix of order 2^{n-1}. Clearly $X(a_{i1},\ldots,a_{in})$ is lower triangular. Also,

$$\operatorname{per}(A)E_{2^n,1} = \prod_{i=1}^{n} X(A_{(i)}). \tag{1.4}$$

ISBN 0-201-13505-1

Example 1.1. If $n=2$, then

$$\operatorname{per}(A)E_{41}=X(A_{(1)})X(A_{(2)})$$

$$=\left[\begin{array}{cc|cc} 0 & 0 & 0 & 0 \\ a_{12} & 0 & 0 & 0 \\ \hline a_{11} & 0 & 0 & 0 \\ 0 & a_{11} & a_{12} & 0 \end{array}\right]\left[\begin{array}{cc|cc} 0 & 0 & 0 & 0 \\ a_{22} & 0 & 0 & 0 \\ \hline a_{21} & 0 & 0 & 0 \\ 0 & a_{21} & a_{22} & 0 \end{array}\right]$$

$$=\left[\begin{array}{ccc|c} 0 & 0 & 0 & 0 \\ \hline 0 & 0 & 0 & 0 \\ 0 & 0 & 0 & 0 \\ a_{11}a_{22}+a_{12}a_{21} & 0 & 0 & 0 \end{array}\right].$$

If $n=3$, then

$$\operatorname{per}(A)E_{81}=X(A_{(1)})X(A_{(2)})X(A_{(3)})$$

$$=\left[\begin{array}{cc|cc|cc|cc} 0 & 0 & 0 & 0 & 0 & 0 & 0 & 0 \\ a_{13} & 0 & 0 & 0 & 0 & 0 & 0 & 0 \\ \hline a_{12} & 0 & 0 & 0 & 0 & 0 & 0 & 0 \\ 0 & a_{12} & a_{13} & 0 & 0 & 0 & 0 & 0 \\ \hline a_{11} & 0 & 0 & 0 & 0 & 0 & 0 & 0 \\ 0 & a_{11} & 0 & 0 & a_{13} & 0 & 0 & 0 \\ \hline 0 & 0 & a_{11} & 0 & a_{12} & 0 & 0 & 0 \\ 0 & 0 & 0 & a_{11} & 0 & a_{12} & a_{13} & 0 \end{array}\right]\cdot X(A_{(2)})X(A_{(3)}).$$

It follows from (1.4) that for any matrix norm we have

$$|\operatorname{per}(A)| \leqslant \prod_{i=1}^{n} \|X(A_{(i)})\|. \tag{1.5}$$

Wilf [unpublished] used inequality (1.5) with the Hilbert (spectral) norm to obtain a slight improvement in Minc's bound [84] for the permanents of (0, 1)-matrices. Jurkat and Ryser [112] used intricate combinatorial manipulations to deduce from (1.4) the following "economy equation," which expresses per(A), the $(2^n, 1)$ entry of the matrix on the left-hand side of (1.4), as a product of blocks of suitably rearranged matrices $X(A_{(i)})$ on the right-hand side of (1.4).

ISBN 0-201-13505-1

THEOREM 1.1 [112]. *If $A=(a_{ij})$ is an $n\times n$ matrix, then*

$$\operatorname{per}(A)=\prod_{i=1}^{n} P_i(A_{(i)}), \tag{1.6}$$

where $P_t(A_{(i)})$ is the $\binom{n}{t-1}\times\binom{n}{t}$ matrix defined by

$$P_1(A_{(i)})=[a_{i1},\ldots,a_{in}], \qquad P_n(A_{(i)})=[a_{in},\ldots,a_{i1}]^{\mathrm{T}},$$

and for $2\leqslant t\leqslant n-1$,

$$P_t(a_{i1},\ldots,a_{in})=\left[\begin{array}{c:c} P_{t-1}(a_{i2},\ldots,a_{in}) & 0 \\ \hdashline a_{i1}I & P_t(a_{i2},\ldots,a_{in}) \end{array}\right].$$

We omit the proof of this result and illustrate it by means of the following example.

Example 1.2. Let $A=(a_{ij})$ be an $n\times n$ matrix. If $n=2$, then

$$\operatorname{per}(A)=[a_{11}a_{12}]\begin{bmatrix} a_{22} \\ a_{21} \end{bmatrix}.$$

If $n=3$, then

$$\operatorname{per}(A)=[a_{11}a_{12}a_{13}]\left[\begin{array}{cc:c} a_{22} & a_{23} & 0 \\ \hdashline a_{21} & 0 & a_{23} \\ 0 & a_{21} & a_{22} \end{array}\right]\begin{bmatrix} a_{33} \\ a_{32} \\ a_{31} \end{bmatrix}.$$

If $n=4$, then

$$\operatorname{per}(A)=[a_{11}a_{12}a_{13}a_{14}]\left[\begin{array}{ccc:ccc} a_{22} & a_{23} & a_{24} & 0 & 0 & 0 \\ \hdashline a_{21} & 0 & 0 & a_{23} & a_{24} & 0 \\ 0 & a_{21} & 0 & a_{22} & 0 & a_{24} \\ 0 & 0 & a_{21} & 0 & a_{22} & a_{23} \end{array}\right]$$

$$\cdot\left[\begin{array}{ccc:c} a_{33} & a_{34} & 0 & 0 \\ a_{32} & 0 & a_{34} & 0 \\ 0 & a_{32} & a_{33} & 0 \\ \hdashline a_{31} & 0 & 0 & a_{34} \\ 0 & a_{31} & 0 & a_{33} \\ 0 & 0 & a_{31} & a_{32} \end{array}\right]\begin{bmatrix} a_{44} \\ a_{43} \\ a_{42} \\ a_{41} \end{bmatrix}.$$

ISBN 0-201-13505-1

The "economy equation" (1.6) yields the following general upper bound:

$$|\mathrm{per}(A)| \leqslant \prod_{i=1}^{n} \|P_i(A_{(i)})\|. \tag{1.7}$$

Jurkat and Ryser [112] exploited the inequality (1.7) to obtain upper bounds for the permanents of (0, 1)-matrices, nonnegative matrices, etc. We shall present some of these in the following sections of this chapter.

6.2 (0, 1)-Matrices

In 1960 Ryser [68] conjectured that in the class of $vk \times vk$ (0, 1)-matrices with row sums and column sums equal to k the permanent function takes its maximum on the direct sum of $k \times k$ matrices of 1's. In 1963 Minc [84] conjectured that, if A is an $n \times n$ (0, 1)-matrix with row sums $r_1, \ldots, r_n$, then

$$\mathrm{per}(A) \leqslant \prod_{i=1}^{n} (r_i!)^{1/r_i}, \tag{2.1}$$

and proved that

$$\mathrm{per}(A) \leqslant \prod_{i=1}^{n} \frac{r_i+1}{2} \tag{2.2}$$

(see Problems 3 and 4). Clearly, Minc's conjecture implies Ryser's conjecture. It also implies the inequality (2.2), since $(r_i!)^{1/r_i}$ is the geometric mean of the first r_i positive integers, while $(r_i+1)/2$ is their arithmetic mean.

Jurkat and Ryser [112], Minc [132], Wilf [147], and Nijenhuis and Wilf [182, 183] obtained various bounds that are weaker versions of (2.1) but sharper than (2.2). Minc [84, 250] proved (2.1) for all matrices whose row sums do not exceed 8. Finally, in 1973 Brégman [219] succeeded in proving Minc's conjecture and thus also Ryser's conjecture. We give below a proof of (2.1) due to Schrijver [303] that is simpler than Brégman's proof. We shall require the following auxiliary results.

Lemma 1. *If $t_1, \ldots, t_r$ are nonnegative real numbers, then*

$$((t_1 + \cdots + t_r)/r)^{t_1 + \cdots + t_r} \leqslant t_1^{t_1} \cdots t_r^{t_r}. \tag{2.3}$$

(The symbol 0^0 is interpreted in this formula as 1.)

The lemma is an immediate consequence of the convexity of the function $x \log x$. For,

$$\left(\frac{t_1 + \cdots + t_r}{r}\right) \log\left(\frac{t_1 + \cdots + t_r}{r}\right) \leqslant \frac{t_1 \log t_1 + \cdots + t_r \log t_r}{r}$$

ISBN 0-201-13505-1

which, after multiplying by r and taking exponents of both sides, yields (2.3).

LEMMA 2. *Let $A=(a_{ij})$ be an $n\times n$ (0,1)-matrix, and let S be the set of all permutations σ such that $\prod_{i=1}^{n} a_{i,\sigma i}=1$. Then*

$$\prod_{i=1}^{n} \prod_{\substack{k \\ a_{ik}=1}} (\operatorname{per} A(i|k))^{\operatorname{per} A(i|k)} = \prod_{\sigma\in S} \prod_{i=1}^{n} \operatorname{per}(A(i|\sigma i)), \tag{2.4}$$

and

$$\prod_{i=1}^{n} r_i^{\operatorname{per} A} = \prod_{\sigma\in S} \prod_{i=1}^{n} r_i. \tag{2.5}$$

Proof. For given i and k, the number of factors $\operatorname{per} A(i|k)$ on the left-hand side of (2.4) is $\operatorname{per} A(i|k)$ if $a_{ik}=1$, and 0 otherwise. The number of such factors on the right-hand side of (2.4) equals the number of permutations σ in S satisfying $\sigma i=k$, which is $\operatorname{per} A(i|k)$ or 0, according as $a_{ik}=1$ or 0.

It is easy to see that for a given i the number of factors r_i on both sides of (2.5) is $\operatorname{per} A$. ■

THEOREM 2.1. *Let A be an $n\times n$ (0,1)-matrix with row sums $r_1,\ldots,r_n$. Then*

$$\operatorname{per}(A) \leqslant \prod_{i=1}^{n} r_i!^{1/r_i}. \tag{2.6}$$

Proof [303]. We use induction on n. By Lemma 1,

$$\begin{aligned}(\operatorname{per} A)^{n \operatorname{per} A} &= \prod_{i=1}^{n} (\operatorname{per} A)^{\operatorname{per} A} \\ &= \prod_{i=1}^{n} \left(\sum_{k=1}^{n} a_{ik} \operatorname{per} A(i|k) \right)^{\Sigma a_{ik} \operatorname{per} A(i|k)} \\ &\leqslant \prod_{i=1}^{n} \left(r_i^{\operatorname{per} A} \prod_{\substack{k \\ a_{ik}=1}} \operatorname{per} A(i|k)^{\operatorname{per} A(i|k)} \right)\end{aligned}$$

and thus, by Lemma 2,

$$(\operatorname{per} A)^{n \operatorname{per} A} \leqslant \prod_{\sigma\in S} \left(\left(\prod_{i=1}^{n} r_i \right) \left(\prod_{i=1}^{n} \operatorname{per} A(i|\sigma i) \right) \right).$$

ISBN 0-201-13505-1

We now apply the induction hypothesis to each $A(i|\sigma i)$:

$$\prod_{i=1}^{n} \operatorname{per} A(i|\sigma i) \leqslant \prod_{i=1}^{n}\left[\prod_{\substack{j\neq i\\ a_{j,\sigma i}=0}} r_j!^{1/r_j}\right]\left[\prod_{\substack{j\neq i\\ a_{j,\sigma i}=1}} (r_j-1)!^{1/(r_j-1)}\right]$$

$$=\prod_{j=1}^{n}\left[\prod_{\substack{i\neq j\\ a_{j,\sigma i}=0}} r_j!^{1/r_j}\right]\left[\prod_{\substack{i\neq j\\ a_{j,\sigma i}=1}} (r_j-1)!^{1/(r_j-1)}\right]$$

$$=\prod_{j=1}^{n} r_j!^{(n-r_j)/r_j}(r_j-1)!^{(r_j-1)/(r_j-1)}.$$

The first equality is just a result of a change in the order of multiplication, and the second equality is obtained by counting the number of factors $r_j!^{1/r_j}$ and factors $(r_j-1)!^{1/(r_j-1)}$. Clearly, for fixed σ and j, the number of i satisfying $i\neq j$ and $a_{j,\sigma i}=0$ is $n-r_j$, and the number of i satisfying $i\neq j$ and $a_{j,\sigma i}=1$ is r_j-1 (since $a_{j,\sigma j}=1$). Hence

$$(\operatorname{per} A)^{n \operatorname{per} A} \leqslant \prod_{\sigma\in S}\left[\left(\prod_{i=1}^{n} r_i\right)\left(\prod_{j=1}^{n} r_j!^{(n-r_j)/r_j}(r_j-1)!\right)\right]$$

$$=\prod_{\sigma\in S}\left(\prod_{i=1}^{n} r_i!^{n/r_i}\right)$$

$$=\left(\prod_{i=1}^{n} r_i!^{1/r_i}\right)^{n \operatorname{per} A},$$

and the result follows. ■

The inequality (2.6) implies Ryser's conjecture. It is not known, however, what is the maximum permanent in the class of $n\times n$ (0, 1)-matrices with k 1's in each row and each column when k does not divide n.

Clearly the bound in (2.6) is not valid for arbitrary matrices with nonnegative integer entries (see Problem 5). Foregger [262] obtained the following bound for the permanents of such matrices.

THEOREM 2.2. *If A is a fully indecomposable $n\times n$ matrix with nonnegative integer entries, then*

$$\operatorname{per}(A) \leqslant 2^{s(A)-2n}+1, \tag{2.7}$$

where $s(A)$ denotes the sum of entries in A.

ISBN 0-201-13505-1

For a proof of the theorem the reader is referred to [262], where conditions for equality in (2.7) are also given.

The bounds in (2.6) and (2.7) are not comparable: If all the row sums of a fully indecomposable (0,1)-matrix are greater than 2, then Theorem 2.1 gives a better bound; if some of the row sums equal 2, then Theorem 2.2 may give a better bound (see Problem 6).

6.3 Nonnegative Matrices

In Section 4.4 we gave lower and upper bounds for permanents of nonnegatve matrices (Theorems 4.1 and 4.2). We quote here these upper bounds again for reference. They can be proved *mutatis mutandis* by the method used in Section 4.4.

If $\alpha=(a_1,\ldots,a_n)$ is a real n-tuple, let $a_1^*\geqslant a_2^*\geqslant\cdots\geqslant a_n^*$ be the numbers $a_1,\ldots,a_n$ arranged in nonincreasing order, and let $a_1'\leqslant a_2'\leqslant\cdots\leqslant a_n'$ be the same numbers arranged in nondecreasing order.

THEOREM 3.1 [112, 161]. *Let $A=(a_{ij})$ be a nonnegative $n\times n$ matrix. Then*

$$\operatorname{per}(A)\leqslant \min_{\sigma\in S_n}\prod_{i=1}^{n}\sum_{t=1}^{i}a_{\sigma i,t}^{*}. \tag{3.1}$$

If A is positive, then equality can occur in (3.1) *if and only if A contains an $(n-1)\times n$ submatrix all of whose rows are multiples of* $(1,1,\ldots,1)$.

For a proof of the condition for equality, see [161].

THEOREM 3.2 [161]. *If $A=(a_{ij})$ is a nonnegative $n\times n$ matrix, then*

$$\operatorname{per}(A)\leqslant\prod_{i=1}^{n}\sum_{k=1}^{i}a_{ik}^{*}-(na_{11}^{*}-r_1)\prod_{h=2}^{n}\sum_{k=1}^{h-1}a_{hk}'.$$

Example 3.1. Let

$$A=\begin{bmatrix}2&1&1\\1&2&1\\1&1&2\end{bmatrix}.$$

Then the bounds given by Theorems 3.1 and 3.2 are 24 and 20, respectively. The permanent of A is actually 16.

If $A=(a_{ij})$ is a nonnegative matrix with row sums $r_1,\ldots,r_n$ and column sums $c_1,\ldots,c_n$, then obviously

$$\operatorname{per}(A)\leqslant\prod_{i=1}^{n}r_i, \tag{3.2}$$

ISBN 0-201-13505-1

and similarly

$$\operatorname{per}(A) \leqslant \prod_{i=1}^{n} c_i. \tag{3.3}$$

Hence

$$\operatorname{per}(A) \leqslant \min\left(\prod_{i=1}^{n} r_i, \prod_{i=1}^{n} c_i \right). \tag{3.4}$$

Jurkat and Ryser [126] obtained the following result, which improves the bound in (3.4).

THEOREM 3.3. *If A is a nonnegative matrix with row sums $r_1 \leqslant r_2 \leqslant \cdots \leqslant r_n$ and column sums $c_1 \leqslant c_2 \leqslant \cdots \leqslant c_n$, then*

$$\operatorname{per}(A) \leqslant \prod_{i=1}^{n} \min(r_i', c_i'). \tag{3.5}$$

We shall require the following lemma: *If $a_1, \ldots, a_n$ and $b_1, b_2, \ldots, b_n$ are nonnegative real numbers, then*

$$\prod_{j=1}^{n} \min(a_j', b_j) \leqslant \prod_{j=1}^{n} \min(a_j', b_j'). \tag{3.6}$$

The lemma is easily provable by induction on n.

Now for the proof of the theorem. We may assume without loss of generality that $r_1 \leqslant \cdots \leqslant r_n$, $c_1 \leqslant \cdots \leqslant c_n$, and that $r_1 \leqslant c_1$. For any j, $1 \leqslant j \leqslant n$, let A_j denote $A(1|j)$, and let the column sums and the row sums of A_j be denoted by $r_2(A_j), \ldots, r_n(A_j)$ and $c_1(A_j), \ldots, c_{j-1}(A_j)$, $c_{j+1}(A_j), \ldots, c_n(A_j)$, respectively. Let $(s_1, s_2, \ldots, s_{n-1})$ be a permutation of $(2, 3, \ldots, n)$ such that

$$r_{s_1}(A_j) \leqslant r_{s_2}(A_j) \leqslant \cdots \leqslant r_{s_{n-1}}(A_j),$$

and let $(t_1, t_2, \ldots, t_{n-1})$ be a permutation of $(1, \ldots, j-1, j+1, \ldots, n)$ such that

$$c_{t_1}(A_j) \leqslant c_{t_2}(A_j) \leqslant \cdots \leqslant c_{t_{n-1}}(A_j).$$

We use induction on n. Expanding the permanent of A by the first row,

ISBN 0-201-13505-1

and using the fact that $r_{s_k}(A_j) \leqslant r_{s_k}$ and $c_{t_k}(A_j) \leqslant c_{t_k}$, for all j, we have

$$\begin{aligned}\operatorname{per}(A) &= \sum_{j=1}^{n} a_{1j} \operatorname{per}(A_j)\\ &\leqslant \sum_{j=1}^{n} a_{1j} \prod_{k=1}^{n-1} \min\big(r_{s_k}(A_j), c_{t_k}(A_j)\big)\\ &\leqslant \sum_{j=1}^{n} a_{1j} \prod_{k=1}^{n-1} \min(r_{s_k}, c_{t_k}).\end{aligned}$$

But by the lemma,

$$\prod_{k=1}^{n-1} \min(r_{s_k}, c_{t_k}) \leqslant \prod_{i=2}^{j} \min(r_i, c_{i-1}) \prod_{i=j+1}^{n} \min(r_i, c_i),$$

and, since $\min(r_i, c_{i-1}) \leqslant \min(r_i, c_i)$ for all i, we have

$$\begin{aligned}\operatorname{per}(A) &\leqslant \sum_{j=1}^{n} a_{1j} \prod_{i=2}^{n} \min(r_i, c_i)\\ &= \prod_{i=1}^{n} \min(r_i, c_i),\end{aligned}$$

since $\sum_{j=1}^{n} a_{1j} = r_1 = \min(r_1, c_1)$. ■

Theorem 3.3 improves the bound in (3.4). It is remarkable that similar improvement is not possible for other bounds. For example,

$$\operatorname{per}(A) \leqslant \min\left(\prod_{i=1}^{n} \frac{r_i + 1}{2}, \prod_{j=1}^{n} \frac{c_j + 1}{2} \right)$$

[see inequality (2.2)]. However,

$$\operatorname{per}(A) \leqslant \prod_{i=1}^{n} \min\left(\frac{r_i' + 1}{2}, \frac{c_i' + 1}{2} \right)$$

is not valid in general (see [126] for a counter-example; also see Problem 12).

ISBN 0-201-13505-1

6.4 Complex Matrices

Several of the upper bounds for the permanents of nonnegative matrices given in the preceding section have their analogues for complex matrices. In fact, since $|\mathrm{per}(A)| \leqslant \mathrm{per}(|A|)$ by the triangle inequality, any such bound can be used to produce an upper bound for the permanents of complex matrices. For example, from inequality (3.4) we obtain

$$|\mathrm{per}(A)| \leqslant \min\left(\prod_{i=1}^{n} p_i, \prod_{j=1}^{n} q_j \right), \tag{4.1}$$

where $p_i = \sum_{s=1}^{n} |a_{is}|$ and $q_j = \sum_{t=1}^{n} |a_{tj}|, i,j = 1, \ldots, n$, and from (3.5) we have

$$|\mathrm{per}(A)| \leqslant \prod_{i=1}^{n} \min(p_i', q_i'). \tag{4.2}$$

Also, the inequality (3.1) yields

$$|\mathrm{per}(A)| \leqslant \min_{\sigma \in S_n} \prod_{i=1}^{n} \sum_{t=1}^{i} |a_{\sigma(i),t}|^*. \tag{4.3}$$

In this section we present more interesting and sophisticated inequalities: upper bounds for the permanents of normal and general complex matrices in terms of their eigenvalues and singular values, respectively. All these results are due to Marcus and Minc [93, 102, 116].

THEOREM 4.1. *If A is an $n \times n$ normal matrix with eigenvalues $\lambda_1, \ldots, \lambda_n$, then*

$$|\mathrm{per}(A)| \leqslant \frac{1}{n} \sum_{i=1}^{n} |\lambda_i|^n. \tag{4.4}$$

For $n > 2$, equality in (4.4) can occur if and only if A is a scalar multiple of a unitary matrix with exactly one nonzero entry in each row and column.

We shall require two preliminary results. We have already encountered the first of these [see formula (5.8), Section 4.5].

LEMMA 1. *If $A = U^*DU$, where U is an $n \times n$ unitary matrix and $D = \mathrm{diag}(\lambda_1, \ldots, \lambda_n)$, then*

$$\mathrm{per}(A) = \sum_{\gamma \in G_{n,n}} \frac{1}{\nu(\gamma)} |\mathrm{per}(U[\gamma|1, \ldots, n])|^2 \prod_{t=1}^{n} \lambda_t^{m_t(\gamma)}, \tag{4.5}$$

ISBN 0-201-13505-1

where $m_t(\gamma)$ *denotes the number of occurrences of* t *in the sequence* $\gamma\in G_{n,n}$, *and* $\nu(\gamma)=\prod_{i=1}^{n} m_t(\gamma)!$.

Formula (4.5) can be used to derive a special relation that holds among the permanents of certain matrices constructed from the rows of a unitary matrix. If we regard $\lambda_1,\ldots,\lambda_n$ as variables and differentiate both sides of (4.5) with respect to λ_s, and evaluate both sides of the resulting equation at $\lambda_t=1, t=1,\ldots,n$, we obtain the following result.

LEMMA 2. *If* U *is an* $n\times n$ *unitary matrix and* s *is an integer*, $1\leqslant s\leqslant n$, *then*

$$\sum_{\gamma\in G_{n,n}} \frac{m_s(\gamma)}{\nu(\gamma)}|\operatorname{per}(U[\gamma|1,\ldots,n])|^2=1. \tag{4.6}$$

Proof of Theorem 4.1. Let $A=U^*DU$, where U is unitary and $D=\operatorname{diag}(\lambda_1,\ldots,\lambda_n)$. For a sequence γ in $G_{n,n}$, let

$$c_\gamma=|\operatorname{per}(U[\gamma|1,\ldots,n])|^2.$$

Then by Lemma 1,

$$\begin{aligned}
|\operatorname{per}(A)| &= \left|\sum_{\gamma\in G_{n,n}} \frac{c_\gamma}{\nu(\gamma)}\prod_{t=1}^{n}\lambda_t^{m_t(\gamma)}\right| \\
&\leqslant \sum_{\gamma\in G_{n,n}} \frac{c_\gamma}{\nu(\gamma)}\prod_{t=1}^{n}|\lambda_t|^{m_t(\gamma)} \\
&\leqslant \sum_{\gamma\in G_{n,n}} \frac{c_\gamma}{\nu(\gamma)}\left(\frac{1}{n}\sum_{t=1}^{n} m_t(\gamma)|\lambda_t|\right)^n \\
&\leqslant \sum_{\gamma\in G_{n,n}} \frac{c_\gamma}{\nu(\gamma)}\frac{1}{n}\sum_{t=1}^{n} m_t(\gamma)|\lambda_t|^n \\
&= \frac{1}{n}\sum_{t=1}^{n}|\lambda_t|^n \sum_{\gamma\in G_{n,n}} \frac{m_t(\gamma)}{\nu(\gamma)}c_\gamma \\
&= \frac{1}{n}\sum_{t=1}^{n}|\lambda_t|^n.
\end{aligned}$$

The first inequality is a consequence of the triangle inequality, the second inequality is the arithmetic–geometric mean inequality, and the third inequality holds by a more general weighted mean theorem (see [44], Theorem 16). The last equality follows from Lemma 2.

For a proof of the case of equality, which is somewhat involved, the reader is referred to [116]. ■

ISBN 0-201-13505-1

If A happens to be positive semi-definite hermitian with eigenvalues $\lambda_1 \geqslant \cdots \geqslant \lambda_n$ and $\omega \in Q_{k,n}$, then it follows immediately from Theorem 4.1 and the Cauchy inequalities (see, e.g., [92], Chapter II, 4.4.7) that

$$\operatorname{per}(A[\omega|\omega]) \leqslant \frac{1}{k} \sum_{j=1}^{k} \lambda_j^k. \tag{4.7}$$

On the other hand, if A is any normal doubly stochastic matrix, then Theorem 4.1 implies that

$$\operatorname{per} A \leqslant \rho(A)/n. \tag{4.8}$$

This is so because the eigenvalues of A do not exceed 1 in modulus, and exactly $\rho(A)$ of them are different from zero.

We can extend the result in Theorem 4.1 to the case of an arbitrary complex square matrix. Recall that, if A is a complex square matrix, then the square roots of the eigenvalues of AA^* are called the *singular values* of A. We combine Theorem 4.1 and Corollary 1 to Theorem 2.4, Chapter 2, to obtain the following upper bound for the permanents of complex matrices that are not necessarily normal.

THEOREM 4.2. *If A is a complex $n \times n$ matrix with singular values $\alpha_1, \ldots, \alpha_n$, then*

$$|\operatorname{per}(A)|^2 \leqslant \frac{1}{n} \sum_{i=1}^{n} \alpha_i^{2n}. \tag{4.9}$$

Equality holds in (4.9) *if and only if $A = DP$, where $D = \operatorname{diag}(d_1, \ldots, d_n)$, $|d_1| = \cdots = |d_n|$, and P is a permutation matrix.*

We leave the discussion of the case of equality as an exercise (Problem 15).

Example 4.1. Show that, if A is an arbitrary $n \times n$ doubly stochastic matrix, then

$$\operatorname{per}(A) \leqslant (\rho(A)/n)^{1/2}. \tag{4.10}$$

By Corollary 1 to Theorem 2.4, Chapter 2, and inequality (4.8) we have

$$\begin{aligned}(\operatorname{per}(A))^2 &\leqslant \operatorname{per} AA^* \\ &\leqslant \rho(AA^*)/n \\ &= \rho(A)/n.\end{aligned}$$

ISBN 0-201-13505-1

Problems

1. Let A_t be an $m_t \times n_t$ matrix for $t=1,\ldots,k$. Show that the row sums of $A_1 \otimes \cdots \otimes A_m$ are

$$\prod_{t=1}^{k} r_{i_t}(A_t),$$

where $r_{i_t}(A_t)$ denotes the i_tth row sum of A_t, and the indices i_t run independently over all integers $1 \leqslant i_t \leqslant m_t, t=1,\ldots,k$.
2. Let $A=(a_{ij})$ be an $n \times n$ matrix. Show that the inequality (1.5) implies

$$|\operatorname{per}(A)| \leqslant \prod_{i=1}^{n} \sum_{j=1}^{n} |a_{ij}|.$$

3. Let $r_1,\ldots,r_c$ be positive integers. Show that [84]:

$$\sum_{j=1}^{c} \frac{2}{r_j} \prod_{t=1}^{c} \frac{r_t}{r_t+1} \leqslant 1,$$

with equality if and only if $c \leqslant 2$ and either r_1 or r_2 is equal to 1.
4. Prove, without using Theorem 2.1, that if A is an $n \times n$ (0,1)-matrix with row sums $r_1,\ldots,r_n$, then

$$\operatorname{per}(A) \leqslant \prod_{i=1}^{n} \frac{r_i+1}{2}.$$

(*Hint*. Expand the permanent by the first column, use induction on n, and apply the inequality in Problem 3.)
5. Show that the bound in Theorem 2.1 is not valid for matrices with nonnegative integer entries.
6. Find a fully indecomposable (0,1)-matrix with permanent greater than 2, for which Theorem 2.2 provides a better bound than that given by Theorem 2.1.
7. Show that the upper bound in Theorem 2.2 does not hold, in general, for partly decomposable matrices.
8. Show that the upper bound in Theorem 2.2 does not hold for arbitrary nonnegative matrices.
9. Prove Theorem 3.1 (cf. Theorem 4.1, Chapter 4).
10. Prove Theorem 3.2 (cf. Theorem 4.2, Chapter 4).
11. Prove the inequality (3.6).
12. Let A denote an $n \times n$ (0,1)-matrix with row sums $r_1,\ldots,r_n$ and column sums $c_1,\ldots,c_n$. Show that the remark at the conclusion of Section 6.3

ISBN 0-201-1350-1

applies to Theorem 2.1—that is, that in general the inequality

$$\operatorname{per}(A) \leqslant \prod_{i=1}^{n} \min\left(r_i'!^{1/r_i'}, c_i'!^{1/c_i'}\right)$$

is not valid.

(*Hint*. Consider the matrix $A=\begin{bmatrix} I & I \\ J & J \end{bmatrix}$, where $I=\begin{bmatrix} 1 & 0 \\ 0 & 1 \end{bmatrix}$ and $J=\begin{bmatrix} 1 & 1 \\ 1 & 1 \end{bmatrix}$.)

13. Compute the upper bounds (4.4), (4.8), and (4.9) for the matrices $I_3, \frac{1}{2}(I_3+P)$, and J_3, and the matrix

$$B=\frac{1}{4}\begin{bmatrix} 0 & 2 & 2 \\ 2 & 1 & 1 \\ 2 & 1 & 1 \end{bmatrix}.$$

Compare the bounds obtained with the actual permanents of the matrices.

14. Let A be a 2×2 normal matrix with eigenvalues λ_1 and λ_2. Find necessary and sufficient conditions that

$$|\operatorname{per}(A)| = \tfrac{1}{2}\left(|\lambda_1|^2 + |\lambda_2|^2\right).$$

15. Prove the condition for equality in Theorem 4.2.
16. Prove Lemma 2, Section 6.4, in detail.

ISBN 0-201-1350-1

CHAPTER 7

Evaluation of Permanents

7.1 Binet–Minc Method

The evaluation of the permanent of an $n\times n$ matrix, whether by direct use of the definition of permanents or by means of Laplace expansion, requires $O((n+1)!)$ multiplications. This large number cannot be pared down by triangularization as in the case of determinants, since elementary operations that leave the determinant of a matrix invariant change its permanent, in general. In this chapter we present evaluation methods that require considerably fewer multiplications than $O((n+1)!)$. The relative merits of these methods will be discussed at the end of the chapter.

Let $A=(a_{ij})$ be an $m\times n$ matrix over any commutative ring. Let $r_{i_1*i_2*\cdots *i_s}$ denote the sum of the entries in the Hadamard product of rows $i_1,\ldots,i_s$ of A; that is,

$$r_{i_1*i_2*\cdots *i_s}=\sum_{j=1}^{n} a_{i_1j}a_{i_2j}\cdots a_{i_sj}.$$

Let $G(m)$ be the set of nondecreasing sequences of positive integers $(t_1,\ldots,t_k)$ such that $t_1+\cdots+t_k=m$. For $(t_1,\ldots,t_k)\in G(m)$, the *S-function* $S(t_1,\ldots,t_k)$ is the symmetrized sum of all distinct products of the $r_{i_1*\cdots *i_s}$, $s=t_1,\ldots,t_k$ so that in each product the sequences $(i_1,\ldots,i_s)\in Q_{s,m}$, $s=t_1,\ldots,t_k$, partition the set $\{1,\ldots,m\}$ (see Section 1.2).

Binet [1] gave the following formulas:

If A is a $2\times n$ matrix, then

$$\operatorname{Per}(A)=S(1,1)-S(2); \tag{1.1}$$

if A is a $3\times n$ matrix, then

$$\operatorname{Per}(A)=S(1,1,1)-S(1,2)+S(3); \tag{1.2}$$

ENCYCLOPEDIA OF MATHEMATICS and Its Applications, Gian-Carlo Rota (ed.).
Vol. 6: Henryk Minc, Permanents

ISBN 0-201-13505-1

if A is a $4\times n$ matrix, then

$$\operatorname{Per}(A)=S(1,1,1,1)-S(1,1,2)+2S(1,3)+S(2,2)-6S(4). \quad (1.3)$$

We now present a recursive formula for the permanents of general $m\times n$ matrices, $m\leqslant n$. The formula is due to Minc [301] and is based to a certain extent on Joachimstal's [5] proof of Binet's formulas.

THEOREM 1.1. *Let* $A=(a_{ij})$ *be an* $m\times n$ *matrix*, $2\leqslant m\leqslant n$. *Then*

$$\operatorname{Per}(A)=\operatorname{Per}(A(m|-))r_m+\sum_{k=1}^{m-1}(-1)^k k!\sum_{\omega\in Q_{k,m-1}}\operatorname{Per}(A(\omega,m|-))r_{\omega*m}, \quad (1.4)$$

where $\operatorname{Per}(A(\omega,m|-))$ *is interpreted as* 1 *for* $\omega=(1,\ldots,m-1)$, *and* $r_{\omega*m}$ *is an abbreviated notation for* $r_{i_1*\cdots*i_k*m}$.

Proof. Each of the terms of the expansion of $\operatorname{Per}(A)$ appears in the first summand on the right-hand side of (1.4), $\operatorname{Per}(A(m|-))r_m$, with coefficient 1, and it does not appear at all in the other summands. The other terms in various summands on the right-hand side of (1.4) are of the form

$$a_{i_1j}a_{i_2j}\cdots a_{i_sj}a_{mj}\prod_{t=s+1}^{m-1}a_{i_tj_t}, \quad (1.5)$$

where $1\leqslant s\leqslant m-1$, $(i_1,\ldots,i_{m-1})$ is a permutation of $(1,\ldots,m-1)$, and $j,j_{s+1},\ldots,j_{m-1}$ are distinct integers in $\{1,\ldots,n\}$. It remains to show that the sum of the coefficients of each term (1.5) in the summands

$$(-1)^k k!\,r_{\omega*m}\operatorname{Per}(A(\omega,m|-)), \quad (1.6)$$

$\omega\in Q_{k,m-1}, k=0,1,\ldots,m-1$, is zero. [For $k=0$ the expression (1.6) is interpreted as $r_m\operatorname{Per}(A(m|-))$.] Each term (1.5) appears exactly once in $r_{\omega*m}\operatorname{Per}(A(\omega,m|-))$ where $\omega=(i_1,\ldots,i_s)\in Q_{s,m-1}$, and exactly once in each of the s summands $r_{\omega*m}\operatorname{Per}(A(\omega,m|-))$ if $\omega\in Q_{s-1,m-1}$ is a subsequence of $(i_1,\ldots,i_s)$. Thus the term (1.5) appears on the right-hand side of (1.4) with coefficient

$$(-1)^s s!+s\big((-1)^{s-1}(s-1)!\big)=0. \qquad \blacksquare$$

Formula (1.4) can be used to derive an explicit formula for the permanent of an $m\times n$ matrix A in terms of the functions $S(\omega)$. Of course, the coefficients in the formula still depend on partitions of m.

THEOREM 1.2 [301]. *Let* A *be an* $m\times n$ *matrix*, $2\leqslant m\leqslant n$. *Then*

$$\operatorname{Per}(A)=\sum_{\omega\in G(m)}c(\omega)S(\omega), \quad (1.7)$$

ISBN 0-201-13505-1

where the coefficient $c(\omega)$ *for* $\omega=(\omega_1,\ldots,\omega_k)$ *in* $G(m)$ *is defined by*

$$c(\omega)=(-1)^{m+k}\prod_{i=1}^{n}(\omega_i-1)!. \tag{1.8}$$

Proof. Formula (1.7) states that Per(A) is a linear combination of the S-functions. This is an immediate consequence of Theorem 1.1, by virtue of the symmetry of the permanent function with respect to rows of the matrix. It remains to prove formula (1.8), which gives the coefficients $c(\omega)$. Let $\omega=(\omega_1,\ldots,\omega_k)\in G(m)$. If $k=1$, then $S(\omega)=r_{1*2*\cdots*m}$ and, by (1.4), $c(\omega)=(-1)^{m-1}(m-1)!$. We now use induction on k. Let $\alpha\in Q_{\omega_1-1,m-1}$, and consider the terms of $r_{\alpha*m}\text{Per}(A(\alpha,m|-))$ [or of $r_m\text{Per}(A(m|-))$ in case $\omega_1=1$] that belong to $S(\omega)$. They are the terms of $r_{\alpha*m}S(\omega_2,\ldots,\omega_k)$ [or of $r_mS(\omega_2,\ldots,\omega_k)$], and they appear with the coefficient $(-1)^{\omega_1-1}(\omega_1-1)!c(\omega_2,\ldots,\omega_k)$ which, by symmetry, is the coefficient of $S(\omega)$ in (1.7). Hence, using the induction hypothesis, we obtain

$$\begin{aligned} c(\omega)&=(-1)^{\omega_1-1}(\omega_1-1)!c(\omega_2,\ldots,\omega_k)\\ &=(-1)^{\omega_1-1}(\omega_1-1)!(-1)^{m-\omega_1+k-1}\prod_{i=2}^{k}(\omega_i-1)!\\ &=(-1)^{m+k}\prod_{i=1}^{k}(\omega_i-1)!. \end{aligned}$$

■

Example 1.1. We obtain an identity relating the coefficients $c(\omega)$ with $|S(\omega)|$, the number of terms in $S(\omega)$ (cf. Problem 3, Chapter 1).

Consider the $m\times m$ matrix A_t all of whose rows consist of t 1's followed by $m-t$ zeros. Then by (1.7),

$$\begin{aligned} \text{per}(A_t)&=\sum_{\omega\in G(m)}c(\omega)S(\omega)\\ &=\sum_{\omega\in G(m)}c(\omega)|S(\omega)|t^{|\omega|}, \end{aligned}$$

where $|\omega|$ denotes the number of integers in the sequence ω. But per(A_t) vanishes for $t=0,1,\ldots,m-1$, and therefore

$$\sum_{\omega\in G(m)}c(\omega)|S(\omega)|t^{|\omega|}=\sum_{k=1}^{m}(-1)^{m-k}E_{m-k}t^k,$$

where E_{m-k} denotes the $(m-k)$th elementary symmetric function of

ISBN 0-201-13505-1

$1,2,\ldots,m-1,\ E_0=1$. It follows that

$$\sum_{\substack{|\omega|=k\\ \omega\in G(m)}} c(\omega)|S(\omega)|=(-1)^{m-k}E_{m-k}.$$

Example 1.2. Compute the permanent of

$$A=\begin{bmatrix} 2 & 1 & 2 & -1 & -2\\ 1 & 2 & 0 & 2 & 0\\ 1 & 1 & -1 & -1 & 1 \end{bmatrix}.$$

Here

$$r_1=2,\ r_2=5,\ r_3=1;$$
$$r_{2*3}=1,\ r_{1*3}=0,\ r_{1*2}=2;$$
$$r_{1*2*3}=6.$$

Therefore

$$S(1,1,1)=10,$$
$$S(1,2)=4,$$
$$S(3)=6.$$

Hence

$$\begin{aligned}\operatorname{Per}(A)&=10-4+2\times 6\\ &=18.\end{aligned}$$

7.2 Ryser's Method

Ryser's formula [87] for the evaluation of permanents is based on the principle of inclusion and exclusion. Essentially, Ryser's method, like the Binet–Minc method, begins with the product of the row sums of the matrix and then discards the superfluous terms in the product. However, the actual procedures of inclusion and exclusion differ in the two methods.

Let A be an $m\times n$ matrix, $m\leqslant n$, and let Λ_k denote the totality of $m\times k$ submatrices of A; that is,

$$\Lambda_k=\{A(-|\beta)|\beta\in Q_{k,n}\}.$$

ISBN 0-201-13505-1

Ryser's formula states [87]:

$$\operatorname{Per}(A)=\sum_{X\in\Lambda_m}\prod_{i=1}^{m}r_i(X)-\binom{n-m+1}{1}\sum_{X\in\Lambda_{m-1}}\prod_{i=1}^{m}r_i(X)+\cdots$$
$$+(-1)^r\binom{n-m+r}{r}\sum_{X\in\Lambda_{m-r}}\prod_{i=1}^{m}r_i(X)+\cdots$$
$$+(-1)^{m-1}\binom{n-1}{m-1}\sum_{X\in\Lambda_1}\prod_{i=1}^{m}r_i(X). \tag{2.1}$$

The proof of (2.1) is quite straightforward: We relegate it to the exercises (see Problem 1).

Ryser's formula (2.1) can be modified so that the first summand is replaced by the product of the row sums of A:

$$\operatorname{Per}(A)=\prod_{i=1}^{m}r_i(A)+\sum_{t=1}^{m-1}c_t\sum_{X\in\Lambda_{m-t}}\prod_{i=1}^{m}r_i(X), \tag{2.2}$$

where $c_1=-1$ and $c_t=-1-\sum_{r=1}^{t-1}\binom{n-m+t}{t-r}c_r$, $t=2,\ldots,m-1$.

We prove the formula (2.2). The first summand on the right-hand side, $\prod_{i=1}^{m}r_i(A)$, contains all the diagonal products of Per(A) and, in addition, products involving two or more entries from the same column. The other summands, $\prod_{i=1}^{m}r_i(X)$, $X\in\Lambda_{m-t}, t=1,\ldots,m-1$, contain only products of the second type, viz., $a_{1\omega_1}a_{2\omega_2}\cdots a_{m\omega_m}$ with $\omega=(\omega_1,\ldots,\omega_m)\in G_{m,n}-Q_{m,n}$. We show that for each such product the sum of coefficients on the right-hand side of (2.2) is 0. Let $\omega=(\omega_1,\ldots,\omega_m)\in G_{m,n}$ be a nondecreasing sequence with exactly k distinct values for the $\omega_i, 1\leqslant k\leqslant m-1$. The product $a_{1\omega_1}a_{2\omega_2}\cdots a_{m\omega_m}$ appears once in $\prod_{i=1}^{m}r_i(A)$ and once in each summand $\prod_{i=1}^{m}r_i(X)$ for which the matrix X contains the k columns of A numbered $\omega_1,\ldots,\omega_m$; there are $\binom{n-k}{m-t-k}$ such matrices X in Λ_{m-t}, $t=1,\ldots,m-k$. Hence the sum of the coefficients of $a_{1\omega_1}a_{2\omega_2}\cdots a_{m\omega_m}$ on the right-hand side of (2.2) is

$$1+\sum_{t=1}^{m-k}\binom{n-k}{m-t-k}c_t.$$

But

$$\begin{aligned}1+\sum_{t=1}^{m-k}\binom{n-k}{m-t-k}c_t&=1+c_{m-k}+\sum_{r=1}^{m-k-1}\binom{n-m+(m-k)}{(m-k)-r}c_r\\&=1+c_{m-k}+(-1-c_{m-k})\\&=0,\end{aligned}$$

by the definition of the c_t in (2.2).

ISBN 0-201-13505-1

If $m=n$, then the formulas (2.1) and (2.2) both become

$$\operatorname{per}(A)=\sum_{t=0}^{n-1}(-1)^t\sum_{X\in\Lambda_{n-t}}\prod_{i=1}^{n}r_i(X). \tag{2.3}$$

Example 2.1. Let A be the matrix in Example 1.2. We evaluate Per(A) using formula (2.1). We compute

$$\sum_{X\in\Lambda_3} r_1(X)r_2(X)r_3(X)=12,$$

$$\sum_{X\in\Lambda_2} r_1(X)r_2(X)r_3(X)=10,$$

$$\sum_{X\in\Lambda_1} r_1(X)r_2(X)r_3(X)=6.$$

Hence

$$\begin{aligned}\operatorname{Per}(A)&=12-3\times 10+6\times 6\\&=18.\end{aligned}$$

Alternatively, using formula (2.2) we obtain:

$$\begin{aligned}\operatorname{Per}(A)&=10-10+3\times 6\\&=18.\end{aligned}$$

It should be stressed that neither the series in formula (2.1) nor that in formula (2.2) converges to Per(A): The sum of the first $m-1$ terms does not approximate, in general, the permanent of A. For instance, in Example 2.1 the sums of the first two terms equal -18 and 0, respectively, whereas the value of the permanent actually is 18.

7.3 Comparison of Evaluation Methods

The efficiency of a particular method for evaluating permanents on computers can be measured by the number of multiplications involved. Thus, the computation of the permanent of an $m\times n$ matrix by a direct application of the definition, with or without the use of a Laplace expansion, requires $(m-1)n!/(n-m)!$ multiplications. For n-square matrices the number is $(n-1)n!$, a prohibitively large number.

Undoubtedly the best method for the evaluation of the permanent of an $n\times n$ matrix A is to find a related $n\times n$ matrix $\tilde{A}$ such that

$$\operatorname{per}(A)=\det(\tilde{A}).$$

ISBN 0-201-13505-1

If such a matrix can be found, then the determinant of $\tilde{A}$, and thus the permanent of A, can be computed in about n^3 multiplications. However, as was shown by Marcus and Minc [70], permanents cannot be transformed uniformly into determinants via a linear transformation; that is, there exists no linear transformation T on $n \times n$ matrices such that

$$\operatorname{per}(A) = \det(T(A))$$

for all $A \in M_n, n \geqslant 3$. Nevertheless, the method may be feasible for special classes of matrices. For example, if $A = (a_{ij})$ is an $n \times n$ Hessenberg matrix (that is, if $a_{ij} = 0$, whenever $j - i \geqslant 2$), then we can construct $\tilde{A} = (\tilde{a}_{ij})$ as follows [153]: Let $\tilde{a}_{ij} = -a_{ij}$ if $j - i = 1$, and $\tilde{a}_{ij} = a_{ij}$ otherwise. Then it is easy to see that $\operatorname{per}(A) = \det(\tilde{A})$ (see Problem 3). In particular, this method applies to Jacobi (tridiagonal) matrices (see Problem 4).

An algorithm for solving perfect matching problems, due to Gal and Breitbart [244], can also be used for computing the permanents of (0, 1)-matrices. The method requires $O(Kn^3)$ steps, where K is the permanent of the matrix, and thus it is efficient for large n provided that the permanent is not large—for example, if the row sums of the matrix do not exceed 3.

Cummings and Wallis [285] devised an algorithm for the evaluation of the permanents of circulants.

For general matrices we have three methods available, together with some variants of them: the Jurkat–Ryser matrix factorization of permanents (Section 6.1), the Binet–Minc method (Section 7.1), and the Ryser method (Section 7.2).

Recall the Jurkat–Ryser economy equation:

$$\operatorname{per}(A) = \prod_{i=1}^{n} P_i(A_{(i)}), \tag{3.1}$$

where A is an $n \times n$ matrix and the $P_i(A_{(i)})$ are the $\binom{n}{i-1} \times \binom{n}{i}$ matrices, $i = 1, \ldots, n$, defined in Theorem 1.1, Chapter 6. The total number of multiplications in the right-hand side of (3.1) is

$$\sum_{i=2}^{n} \binom{n}{i-1}\binom{n}{i} = \binom{2n}{n-1} - n, \tag{3.2}$$

that is, a number of the order 4^n. Probably the labor could be somewhat reduced by taking into acount the special structure of the $P_i(A_{(i)})$ with their many zeros in fixed positions. However, the method is quite unsuitable for computers, owing to the large sizes of the matrices $P_i(A_{(i)})$ requiring impossibly large computer memories. For example, the economy equation for a 15×15 matrix involves 6435×6435 matrices, and that for a 25×25 matrix involves 5200300-square matrices.

ISBN 0-201-13505-1

The Binet–Minc formula, Theorem 1.2, for the permanent of an $n\times n$ matrix involves more than

$$(n-1)\sum_{\omega\in G(n)}|S(\omega)|$$

multiplications (for notation see Section 7.1; in particular, Example 1.1), since each $S(\omega)$ involves at least $n-1$ multiplications. Now,

$$\sum_{\omega\in G(n)}|S(\omega)|$$

equals the number of ways of partitioning the set $\{1,2,\ldots,n\}$ into subsets, that is equal to the sum of Stirling numbers of the second kind, $\Sigma_{r=1}^{n}S(n,r)$, which exceeds $(n/2)^{n/2}$. Thus, the Binet–Minc method requires a large number of multiplications and is not suitable for the computation of permanents on computers.

Ryser's formula

$$\operatorname{per}(A)=\sum_{t=0}^{n-1}(-1)^t\sum_{X\in\Lambda_{n-t}}\prod_{i=1}^{n}r_i(X) \tag{3.3}$$

involves $(n-1)(2^n-1)$ multiplications. For, each of the 2^n-1 nonempty subsets $\{i_1,\ldots,i_k\}$ of $\{1,\ldots,n\}$ corresponds to exactly one matrix X on the right-hand side of (3.3)—namely, $A[1,\ldots,n|i_1,\ldots,i_k]$—and for each such matrix $n-1$ multiplications of the $r_i(X)$ are required. Nijenhuis and Wilf [274] devised a method which further reduces the number of multiplications by a factor of 2, and the number of additions by a factor of $n/2$. They also gave a detailed program in FORTRAN for the evaluation of permanents by the modified Ryser's method.

To summarize: Apart from an *ad hoc* transformation of permanents into determinants and methods applicable to special classes of matrices, the most efficient method for evaluating permanents on computers is the Ryser method.

Problems

1. Prove formula (2.1).
2. Estimate the number of multiplications involved in evaluating the permanent of an $m\times n$ matrix by formula (2.2).
3. Let $A=(a_{ij})$ be an $n\times n$ Hessenberg matrix, and let $\tilde{A}=(\tilde{a}_{ij})$ be the $n\times n$ matrix with $\tilde{a}_{i,i+1}=-a_{i,i+1}, i=1,\ldots,n-1$, and $\tilde{a}_{ij}=a_{ij}$ otherwise. Prove that $\operatorname{per}(A)=\det(\tilde{A})$.
4. Let A be a Jacobi (tridiagonal) matrix, and let $\hat{A}=(\hat{a}_{ij})$ be defined by

ISBN 0-201-13505-1

$\hat{a}_{st} = ia_{st}$ if $s \neq t$ and $\hat{a}_{ss} = a_{ss}$, for all s and t ($i = \sqrt{-1}$). Prove that $\operatorname{per}(A) = \det(\hat{A})$.

5. Let $A = (a_{ij})$ be an $n \times n$ matrix all of whose entries outside the first row, the first column, and the main diagonal are 0 (that is, $a_{ij} \neq 0$ implies that $i = 1$, or $j = 1$, or $i = j$); in addition $a_{11} = 0$. Show that $\operatorname{per}(A) = -\det(A)$.
6. Let A be a 30×30 (0, 1)-matrix with three 1's in each row. Use Theorem 2.1, Chapter 6, to estimate the number of steps required for the computation of per(A) by the Gal–Breitbart algorithm. Compare it with the number required by the Ryser method.
7. Let

$$A = \begin{bmatrix} 3 & -4 & 1 & 2 & 1 \\ 1 & 1 & 0 & 3 & 2 \\ 1 & 2 & -3 & 0 & 4 \end{bmatrix}.$$

   Compute Per(A) using: (a) Theorem 1.2; (b) formula (2.1); (c) formula (2.2).
8. How many operations (multiplications and additions) are involved in computing the permanent of a 3×5 matrix using: (a) the definition of a permanent; (b) Theorem 1.2; (c) formula (2.1); (d) formula (2.2)?

ISBN 0-201-13505-1

CHAPTER 8

More about Permanents

8.1 Other Results

In the first section of this, the concluding chapter of the monograph, we shall deal with some topics in the theory of permanents that do not fall under the headings of the preceding chapters, yet have created considerable interest and have been studied extensively.

Nearly one-tenth of the papers listed in our bibliography deal specifically with subpermanents—that is, permanents of submatrices. Apart from the permanental counterparts of classical results on determinants, such as the Laplace expansion and the Binet–Cauchy theorem for permanents (Theorems 1.2 and 1.3, Chapter 2), subpermanents appear mainly in:

•results involving the set of all subpermanents of a given order of a matrix or, in particular, their sum;

•results involving principal subpermanents of a given order or their sum;

•results on permanents of partitioned matrices, induced matrices, permanental adjugate matrices, etc.

The sum of all subpermanents of order k of an $n\times n$ matrix A, denoted by $\sigma_k(A)$, was studied first by Tverberg [88], who conjectured that, if $A\in\Omega_n$, then

$$\sigma_t(A)\geqslant\sigma_t(J_n), \tag{1.1}$$

$t=1,\ldots,n$, and proved it for $n=2$ and 3. The inequalities (1.1), which generalize the van der Waerden conjecture, were proved by Sasser and Slater [134] for the case when A is a normal doubly stochastic matrix whose spectrum lies in the sector $[-\pi/2t,\pi/2t]$ of the complex plane. This result was sharpened by Marcus and Minc [144], who proved also that, if A

ENCYCLOPEDIA OF MATHEMATICS and Its Applications, Gian-Carlo Rota (ed.). Vol. 6: Henryk Minc, Permanents

ISBN 0-201-13505-1

is a matrix satisfying the conditions of the Sasser–Slater theorem, then

$$\sigma_t(A) \geqslant \frac{(n-t+1)^2}{nt}\sigma_{t-1}(A), \tag{1.2}$$

and in particular

$$\frac{1}{n^2}\sigma_{n-1}(A) \leqslant \operatorname{per}(A). \tag{1.3}$$

Equality can hold in (1.2) if and only if $A = J_n$. The inequality (1.2) confirms, for the special class of matrices described above, an unresolved conjecture of Đoković [123] asserting that (1.2) holds for every $A \in \Omega_n$, with equality if and only if $A = J_n$.

The significance of the inequality (1.2) lies in the fact that

$$\frac{d}{d\theta}\sigma_t(\theta J_n + (1-\theta)A)|_{\theta=0} = \frac{(n-t+1)^2}{n}\sigma_{t-1}(A) - t\sigma_t(A).$$

For,

$$\begin{aligned}\frac{d}{d\theta}\sigma_t(\theta J_n + (1-\theta)A)|_{\theta=0} &= \sum_{\alpha,\beta \in Q_{t,n}} \sum_{\substack{i \in \alpha \\ j \in \beta}} \left(\frac{1}{n} - a_{ij}\right)\operatorname{per}((A[\alpha|\beta])(i|j)) \\ &= \frac{1}{n}\sum_{\alpha,\beta \in Q_{t,n}} \sum_{\substack{i \in \alpha \\ j \in \beta}} \operatorname{per}((A[\alpha|\beta])(i|j)) - t\sigma_t(A) \\ &= \frac{(n-t+1)^2}{n}\sigma_{t-1}(A) - t\sigma_t(A).\end{aligned}$$

Thus, the inequality (1.2) states that the derivative of $\sigma_t(\theta J_n + (1-\theta)A)$ is negative at $\theta = 0$—that is, that the function σ_t decreases at A on the segment joining A to J_n. Hence the Đoković conjecture implies the van der Waerden conjecture. By Theorem 1.4, Chapter 5, the Đoković conjecture holds also for all $A \in \Omega_n$ in a sufficiently small neighborhood of J_n.

An interesting result related to the van der Waerden conjecture was obtained by Sasser and Slater [168], who proved that the only local extremum of $\sigma_t(A)$ [and thus in particular of per(A)] in the interior of Ω_n is at J_n.

Another variation on the theme of the van der Waerden conjecture is the problem of characterizing doubly stochastic matrices all of whose subpermanents of a specified order are equal. We saw (Problem 18, Chapter 5) that, if we could prove that in each minimizing matrix every permanental

ISBN 0-201-13505-1

minor is equal to the permanent of the matrix, then we could conclude that the permanent actually equals $\operatorname{per}(J_n)$. The conclusion is not true if the condition holds merely for a particular doubly stochastic matrix A. For example, all the permanental minors of $\frac{1}{2}(I_n+P)$ equal $\operatorname{per}(\frac{1}{2}(I_n+P))=1/2^{n-1}$ which is greater than $\operatorname{per}(J_n)$ for $n\geqslant 3$. However, Minc [270] obtained the following result.

THEOREM 1.1. *Let $A\in\Omega_n$. If, for some k, $1\leqslant k\leqslant n-2$, all subpermanents of A of order k are equal, then $A=J_n$.*

Proof. We first show that, if all subpermanents of A equal c and if k divides n, then

$$\operatorname{per}(A)=f(n,k,c)=n!\left(\frac{c}{k!}\right)^{n/k}. \tag{1.4}$$

We use induction on n/k. Expanding the permanent of A by the first k rows, we get

$$\begin{aligned}\operatorname{per}(A)&=\sum_{\omega\in Q_{k,n}}\operatorname{per}(A[1,\ldots,k|\omega])\operatorname{per}(A(1,\ldots,k|\omega))\\&=\sum_{\omega\in Q_{k,n}}cf(n-k,k,c)\\&=\binom{n}{k}c(n-k)!\left(\frac{c}{k!}\right)^{(n-k)/k}\\&=n!\left(\frac{c}{k!}\right)^{n/k}\\&=f(n,k,c).\end{aligned}$$

Now, let $n\equiv m \mod k$, $2\leqslant m\leqslant k+1$, and let $\alpha=(i_1,\ldots,i_{m-1})$ and $\beta=(j_1,\ldots,j_{m-1})$ be sequences in $Q_{m-1,n}$. If s is any integer, $1\leqslant s\leqslant n$, not in α, then

$$\begin{aligned}\operatorname{per}(A(\alpha|\beta))&=\sum_{t\notin\beta}a_{st}\operatorname{per}(A(i_1,\ldots,i_{m-1},s|j_1,\ldots,j_{m-1},t))\\&=f(n-m,k,c)\sum_{t\notin\beta}a_{st}\\&=f(n-m,k,c)\left(1-\sum_{r=1}^{m-1}a_{sj_r}\right).\end{aligned} \tag{1.5}$$

Let s_1 and s_2 be any two distinct integers in the complement of α in

ISBN 0-201-13505-1

$\{1,\ldots,n\}$. The formula (1.5) implies that

$$\sum_{r=1}^{m-1} a_{s_1 j_r} = \sum_{r=1}^{m-1} a_{s_2 j_r}.$$

But α is arbitrary. It follows that $\sum_{r=1}^{m-1} a_{sj_r}$ is constant for all s, $1 \leqslant s \leqslant n$; that is,

$$\sum_{r=1}^{m-1} A^{(j_r)} = d_\beta \varepsilon$$

where $\varepsilon = (1,1,\ldots,1)$ and d_β is a positive number depending on $\beta = (j_1,\ldots,j_{m-1})$ only. Now, if we replace one of the j_r in β, say j_1, by t, $t \notin \beta$, we have

$$\sum_{r=2}^{m-1} A^{(j_r)} + A^{(t)} = d_\gamma \varepsilon$$

and therefore

$$A^{(j_1)} - A^{(t)} = (d_\beta - d_\gamma)\varepsilon. \tag{1.6}$$

Since β is arbitrary the equality (1.6) holds for all t, $t \neq j_1$. Hence

$$nA^{(j_1)} - \sum_{t=1}^{n} A^{(t)} = d\varepsilon,$$

where d is some number. But $\sum_{t=1}^{n} A^{(t)} = \varepsilon$, and therefore $A^{(j_1)}$ is a multiple of ε, and since j_1 can be any integer, $1 \leqslant j_1 \leqslant n$, and A is a doubly stochastic matrix, we must have

$$A = J_n. \qquad \blacksquare$$

A similar result for $m \times n$ (0,1)-matrices was obtained by Brualdi and Foregger [260]. They conjectured also that, *if the permanental minors of order $n-1$ of an $n \times n$ (0,1)-matrix A have a common nonzero value, then $A = J$ or $A = I_n + P$*, and they showed that the conjecture holds in certain cases. Sinkhorn [286] proposed a conjecture: *If $A \in \Omega_n$ and* $\text{per}(A(i|j)) \geqslant \text{per}(A)$ *for all i and j, then either $A = J_n$, or $A = \frac{1}{2}(I_n + P)$.* The conjecture, which implies, via Theorem 2.3, Chapter 5, the van der Waerden conjecture, seems to be out of reach.

Brualdi and Newman [97, 109] obtained an upper bound for $\sigma_t(A)$, where A is a substochastic $n \times n$ matrix (that is, a nonnegative matrix A whose

ISBN 0-201-13505-1

row sums do not exceed 1):

$$\sigma_t(A) \leqslant \binom{n}{t}. \tag{1.7}$$

For general complex matrices, Marcus and Gordon [91] obtained upper bounds for the sum of the squares of absolute values of all subpermanents, of order k.

Marcus [115] extended the Marcus–Newman inequality (Theorem 2.4(b), Chapter 2) to direct products: *If A is an $n\times n$ matrix and B is an $m\times m$ matrix, then*

$$|\operatorname{per}(A\otimes B)|^2 \leqslant (\operatorname{per}(AA^*))^m(\operatorname{per}(B^*B))^n. \tag{1.8}$$

Brualdi [107] obtained bounds for $\operatorname{per}(A\otimes B)$ in case A and B are nonnegative. Marcus [158] also generalized the Marcus–Newman inequality as follows:

$$|\sigma_t(AB)|^2 \leqslant \sigma_t(AA^*)\,\sigma_t(B^*B). \tag{1.9}$$

We now turn to results on permanents of principal submatrices. Marcus [82] obtained bounds for the permanent of a positive semi-definite hermitian $n\times n$ matrix $A=(a_{ij})$ in terms of its principal subpermanents:

$$na_{ii}\operatorname{per}(A(i|i)) \geqslant \operatorname{per}(A) \geqslant a_{ii}\operatorname{per}(A(i|i)), \tag{1.10}$$

$i=1,\ldots,n$. The lower bound implies immediately the permanent counterpart of the Hadamard theorem for determinants:

$$\operatorname{per}(A) \geqslant \prod_{i=1}^{n} a_{ii}. \tag{1.11}$$

In [100] Marcus obtained a lower bound for the permanent of $A[\omega]$, where A is positive definite hermitian and $\omega\in G_{k,n}$ (not necessarily in $Q_{k,n}$!), and in [180] he proved that, *if A and B are commuting $n\times n$ positive definite hermitian matrices, $\alpha\in Q_{k,n}$, and $1\leqslant r\leqslant k\leqslant n$, then*

$$\rho_r\big((A+B)^{1/r}[\alpha]\big) \geqslant \rho_r\big(A^{1/r}[\alpha]\big)+\rho_r\big(B^{1/r}[\alpha]\big), \tag{1.12}$$

where $\rho_r(X)$ denotes the sum of all principal subpermanents of X of order r.

Đoković [124] generalized the classical Marcus–Newman result (Corollary 1, Section 4.5): *If $A\in\Omega_n$ is a positive semi-definite matrix and $\alpha\in G_{k,n}$, then*

$$\operatorname{per}(A[\alpha]) \geqslant r!/n^r. \tag{1.13}$$

ISBN 0-201-13505-1

Friedland [243] extended the inequality (1.13) to all matrices in Ω_n whose numerical range lies in the sector $[-\pi/2r, \pi/2r]$ (Theorem 3.6, Chapter 5). Brualdi and Newman [97] proved that, *if* $A \in \Omega_n$, *then*

$$\sum_{i=1}^{n} (1-a_{ii})\operatorname{per}(A(i|i)) \leqslant 1-\operatorname{per}(A). \tag{1.14}$$

Marcus and Minc [102] obtained the following bounds for a principal supermanent of a positive semi-definite hermitian matrix in terms of its eigenvalues:

THEOREM 1.2. *If A is an $n\times n$ positive semi-definite hermitian matrix with eigenvalues $\lambda_1 \geqslant \cdots \geqslant \lambda_n$ and $\alpha \in Q_{r,n}$, then*

$$\prod_{j=1}^{r} \lambda_{n-j+1} \leqslant \operatorname{per}(A[\alpha]) \leqslant \frac{1}{r}\sum_{j=1}^{r} \lambda_j^r. \tag{1.15}$$

Proof. By Schur's theorem (Theorem 2.5, Chapter 2),

$$\operatorname{per}(A[\alpha]) \geqslant \det(A[\alpha]). \tag{1.16}$$

Let $e_1,\ldots,e_n$ be the standard basis of the space of n-tuples. Then $(A[\alpha])_{ij} = (Ae_{\alpha_j}, e_{\alpha_i}) = (Ae_{\alpha_i}, e_{\alpha_j})$. Hence

$$\begin{aligned}\det(A[\alpha]) &= \left(C_r(A)e_{\alpha_1}\wedge\cdots\wedge e_{\alpha_r}, e_{\alpha_1}\wedge\cdots\wedge e_{\alpha_r}\right)\\ &\geqslant \prod_{j=1}^{r}\lambda_{n-j+1},\end{aligned}$$

since $\prod_{j=1}^{r}\lambda_{n-j+1}$ is the smallest eigenvalue of the compound matrix $C_r(A)$. This proves the lower bound. To prove the upper bound we apply Theorem 4.1, Chapter 6, to matrix $A[\alpha]$:

$$\operatorname{per}(A[\alpha]) \leqslant \frac{1}{r}\sum_{i=1}^{r}\mu_i^r,$$

where $\mu_1 \geqslant \cdots \geqslant \mu_r$ are the eigenvalues of $A[\alpha]$. The result follows, since by the Cauchy inequalities (see, e.g., [92; Chapter 2, 4.4.7]), $\mu_i \leqslant \lambda_i$, $i = 1,\ldots,r$. ■

Results on permanents of partitioned matrices will be discussed in detail in Section 8.3 (Conjectures 8 and 9, Theorem 3.2).

ISBN 0-201-13505-1

We conclude the section with some results involving both determinants and permanents. Muir [14,27,30] obtained several relations between determinants and permanents. We quote his result in [18]: *If A is an $n\times n$ matrix, then*

$$\sum_{r=0}^{n}\sum_{\omega\in Q_{r,n}}(-1)^r\operatorname{per}(A[\omega|\omega])\det(A(\omega|\omega))=0,$$

where $\operatorname{per}(A[\omega|\omega])$ *and* $\det(A(\omega|\omega))$ *are interpreted as* 1 *and* $\det(A)$, *respectively, if $r=0$, and as* $\operatorname{per}(A)$ *and* 1 *if $r=n$.*

In [20] Muir expressed the determinant of an $mn\times mn$ matrix A as a sum of permanents of $m\times m$ matrices whose entries are subdeterminants of A. Marcus [74] obtained a lower bound for the determinant of a positive definite hermitian matrix in terms of the product of its principal subpermanents.

Marcus and Merris [234] showed that, *if A is a positive semi-definite hermitian $n\times n$ matrix, and $D(A)$ and $P(A)$ are $n\times n$ matrices whose (i,j) entries are* $\operatorname{per}(A(j|i))$ *and* $(-1)^{i+j}\det(A(j|i))$, *respectively, $i,j=1,\ldots,n$, then*

$$n(\det(A))^{-1}D(A)-(\operatorname{per}(A))^{-1}P(A)$$

is positive semi-definite.

8.2 Some Applications of Permanents

Permanents cannot compete with determinants for preponderance of applications. Determinants are ubiquitous in pure and applied mathematics and in the sciences. Permanents are not so intimately connected to the fundamental properties of matrices, yet they have many uses, particularly in combinatorics and in applied problems of enumerative nature. Even permanents of doubly stochastic matrices and the van der Waerden conjecture, an ostensibly pure intellectual speculation, have applications in probability, combinatorics, and statistical mechanics (*vide infra*).

L. H. Harper (see [147]) suggested the following probabilistic interpretation of the permanent of a doubly stochastic $n\times n$ matrix. Let each of n boxes contain one ball, and let the probability that the ball in box i moves to box j be a_{ij}. Then the probability that after a simultaneous transititon of the balls there is still exactly one ball in each box is $\operatorname{per}((a_{ij}))$.

Some applications of permanents in probability theory were developed by Gyires [227]. He also obtained equalities and inequalities for permanents starting from probabilistic models.

Two of the best-known classical combinatorial problems are the enumerations of Latin rectangles and squares and of nonisomorphic Steiner triple

ISBN 0-201-13505-1

systems of a given order. In both cases the known lower bounds could be considerably improved if the van der Waerden conjecture could be assumed to be true, or if a comparable lower bound could be established for the permanents of (0,1)-matrices in Λ_n^k.

Let $L(r,n)$ denote the number of $r\times n$ Latin rectangles, $r\leqslant n$, based on an n-set S. Let A be a $t\times n$ Latin rectangle based on S, $1\leqslant t<r$, and let S_j be the set of elements of S that do not appear in column j of A. Construct the incidence matrix A' for the subsets $S_1,\dots,S_n$. Clearly, $\operatorname{per}(A')$ is the number of ways that A can be extended to a $(t+1)\times n$ Latin rectangle. Now, $A'\in\Lambda_n^{n-t}$. Let $m(k,n)$ and $M(k,n)$ denote any lower and upper bounds, respectively, for the permanent function in Λ_n^k. Then

$$m(n-t,n)\leqslant \operatorname{per}(A')\leqslant M(n-t,n),$$

$t=1,\dots,r-1$, and thus

$$n!\prod_{t=1}^{r-1} m(n-t,n)\leqslant L(r,n)\leqslant n!\prod_{t=1}^{r-1} M(n-t,n). \tag{2.1}$$

For example, using Theorem 2.1, Chapter 6, we have

$$L(r,n)\leqslant n!\prod_{t=1}^{r-1}(n-t)!^{n/(n-t)}. \tag{2.2}$$

Similarly, we can use any of the lower bounds in Section 4.2 to obtain a lower bound for $L(r,n)$. Of course, a better lower bound for $L(r,n)$ is obtained by the use of the van der Waerden conjecture:

$$L(r,n)\geqslant \frac{n!^r}{n^{n(r-1)}}\prod_{t=1}^{r-1}(n-t)^n, \tag{2.3}$$

since for a given t, $1\leqslant t<r$, the matrix $A'/(n-t)$ is doubly stochastic, and thus applying the conjecture we have

$$\operatorname{per}(A')\geqslant \frac{n!}{n^n}(n-t)^n.$$

The bounds in (2.1), (2.2), and (2.3) can be improved by using the known value of $\operatorname{per}(A')$ for $t=1$ instead of $M(n-1,n)$ and $m(n-1,n)$. Then (2.1) becomes

$$n!d_n\prod_{t=2}^{r-1} m(n-t,n)\leqslant L(r,n)\leqslant n!d_n\prod_{t=2}^{r-1} M(n-t,n), \tag{2.4}$$

where $d_n=n!\sum_{i=0}^{n}(-1)^i/i!$ [see formula (4.2), Section 3.4]. Bounds for the

ISBN 0-201-13505-1

number of Latin squares of order n are obtained from the above inequalities by setting $r=n$. For example, if the van der Waerden conjecture were known to be true, then

$$L(n,n) \geqslant n!^{2n}/n^{n^2}. \tag{2.5}$$

Example 2.1. We use Theorem 1.2, Chapter 4, and formula (2.1) to obtain the following lower bound for the number of Latin squares of order n:

$$L(n,n) \geqslant \prod_{t=1}^{n} t!. \tag{2.6}$$

This bound can be somewhat simplified [259] as follows. Since

$$e^n > 1 + \frac{n}{1!} + \frac{n^2}{2!} + \cdots + \frac{n^t}{t!} > \frac{n^t}{t!}$$

and thus

$$t! > e^{-n} n^t,$$

we have

$$\begin{aligned} L(n,n) &> \prod_{t=1}^{n} (e^{-n} n^t) \\ &= e^{-n^2} n^{1+2+\cdots+n} \\ &= e^{-n^2} n^{n(n+1)/2}, \end{aligned}$$

and therefore

$$L(n,n) > (e^{-2} n)^{n^2/2}. \tag{2.7}$$

Example 2.2. We use the same technique as in the preceding example to simplify formula (2.5):

$$\begin{aligned} L(n,n) &\geqslant n!^{2n}/n^{n^2} \\ &> (e^{-n} n^n)^{2n}/n^{n^2} \\ &= (e^{-2} n)^{n^2}. \end{aligned} \tag{2.8}$$

A *Steiner triple system* of order v, $v \geqslant 3$, is a set of triples of elements of a v-set S such that each pair of elements of S is a subset of exactly one

ISBN 0-201-13505-1

triple. It is not difficult to show that the incidence matrix of a Steiner triple system is a $b\times v$ (0, 1)-matrix A with three 1's in each row and r 1's in each column, where $r=(v-1)/2$ and $b=v(v-1)/6$, and satisfying $AA^{\mathrm{T}}=2I_n+J$. It follows that a Steiner triple system can exist only if $v\equiv 1$ or 3 (mod 6). It is known that this condition is also sufficient. Two Steiner triple systems are isomorphic if their incidence matrices are equal modulo permutations of rows and columns. Let $N(v)$ denote the number of nonisomorphic Steiner triple systems of order v. Wilson [259] used the lower bound in (2.7) to obtain a lower bound for the number of nonisomorphic Steiner triples:

$$N(v)\geqslant(e^{-5}v)^{v^2/12}. \tag{2.9}$$

He also showed that, if the van der Waerden conjecture is assumed to be valid—that is, if the inequality (2.7) can be replaced by the inequality (2.8)—then the bound in (2.9) can be improved to read:

$$N(v)\geqslant(e^{-5}v)^{v^2/6}.$$

Some of the most important applications of permanents are via graph theory. They essentially involve enumerations of certain subgraphs of a graph or a directed graph.

Let D be a directed graph with n vertices $v_1,\ldots,v_n$. A *linear subgraph* of D is a spanning subgraph (that is, one containing all the n vertices) of D such that each v_i has indegree 1 and outdegree 1. In other words, a linear subgraph is a set of disjoint directed cycles. The problem is to find the number of linear subgraphs of D. For this purpose we construct the *adjacency matrix* of D, the $n\times n$ (0, 1)-matrix whose (i,j) entry is 1 if D contains an arc from v_i to v_j, and 0 otherwise. Denote the matrix by $A(D)=(a_{ij})$. Since the indegree of v_i equals the ith column sum and the outdegree of v_i is the ith row sum of D, it follows that to each linear subgraph correspond n positive entries in $A(D)$, one in each row and each column—that is, a positive diagonal of $A(D)$. Hence the number of linear subgraphs of D is the number of positive diagonals in $A(D)$—that is, the permanent of $A(D)$ [154].

A more difficult problem with many applications is the enumeration of the 1-factors of a graph. A *1-factor* of a graph G with $2m$ vertices is a spanning subgraph of G in which every vertex has degree 1. Let $A(G)=(a_{ij})$ be the adjacency matrix of G, and let $\mu(G)$ denote the number of 1-factors of G. Each 1-factor of G is represented by a diagonal of $A(G)$ corresponding to a permutation which is a product of m disjoint transpositions. Therefore the number of 1-factors of G is

$$\mu(G)=\sum_{\sigma\in T}\prod_{i=1}^{2m}a_{i\sigma(i)}, \tag{2.10}$$

ISBN 0-201-13505-1

where the summation is over T, the set of $(2m)!/(2^m m!)$ permutations which can be factored into products of m disjoint transpositions. The function on the right-hand side of (2.10) is called the *hafnian* of $A(G)$. Hafnians are useful matrix functions introduced by Caianiello [54,57]. They are related to pfaffians in the same way as permanents are to determinants. For example, if $C=(c_{ij})$ is a 4×4 matrix, then

$$\operatorname{haf}(C)=c_{12}c_{21}c_{34}c_{43}+c_{13}c_{24}c_{31}c_{42}+c_{14}c_{23}c_{32}c_{41}.$$

If C happens to be an adjacency matrix, then it is symmetric, and

$$\operatorname{haf}(C)=c_{12}^2c_{34}^2+c_{13}^2c_{24}^2+c_{14}^2c_{23}^2.$$

From (2.10) we have

$$\mu(G)=\operatorname{haf}(A(G)). \tag{2.11}$$

Unfortunately, there is no known efficient method for computing hafnians. Gibson [176] observed that the square of the hafnian of a symmetric nonnegative matrix does not exceed its permanent, and therefore

$$\mu(G)\leqslant\sqrt{\operatorname{per}(A(G))}\ . \tag{2.12}$$

Of course, the evaluation of a permanent of a large matrix presents considerable difficulties. However, we can combine the inequality (2.12) with any upper bound for the permanents of (0, 1)-matrices to obtain an upper bound for $\mu(G)$. Thus using Theorem 2.1, Chapter 6, we have

$$\mu(G)\leqslant\prod_{i=1}^{2m} r_i!^{1/2r_i}, \tag{2.13}$$

where r_i is the ith row sum of $A(G)$—that is, the degree of the vertex v_i, $i=1,\dots,2m$.

Example 2.3. Let G be the graph

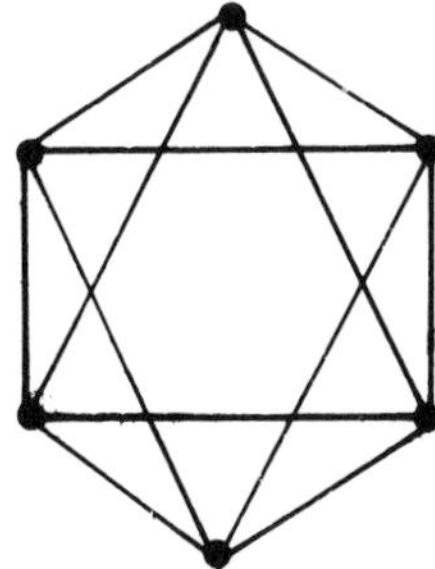

ISBN 0-201-13505-1

Here $m=3, r_i=4$ for $i=1,\ldots,6$, and therefore by (2.13)

$$\mu(G)\leqslant 10.84\ldots;$$

that is,

$$\mu(G)\leqslant 10.$$

Actually

$$\mu(G)=8,$$

since

$$A(G)=\begin{bmatrix} 0 & 1 & 1 & 0 & 1 & 1 \\ 1 & 0 & 1 & 1 & 0 & 1 \\ 1 & 1 & 0 & 1 & 1 & 0 \\ 0 & 1 & 1 & 0 & 1 & 1 \\ 1 & 0 & 1 & 1 & 0 & 1 \\ 1 & 1 & 0 & 1 & 1 & 0 \end{bmatrix}$$

and

$$\begin{aligned}\operatorname{haf}(A(G)) &= a_{12}^2a_{34}^2a_{56}^2 + a_{12}^2a_{35}^2a_{46}^2 + a_{13}^2a_{24}^2a_{56}^2 \\ &\quad + a_{13}^2a_{26}^2a_{45}^2 + a_{15}^2a_{23}^2a_{46}^2 + a_{15}^2a_{26}^2a_{34}^2 \\ &\quad + a_{16}^2a_{23}^2a_{45}^2 + a_{16}^2a_{24}^2a_{35}^2,\end{aligned}$$

the other seven diagonal products of the hafnian being zero. Hence

$$\operatorname{haf}(A(G))=8.$$

We also note that

$$\operatorname{per}(A(G))=80,$$

and therefore the inequality (2.12) gives

$$\begin{aligned}\mu(G) &\leqslant \sqrt{80} \\ &= 8.94\ldots;\end{aligned}$$

that is,

$$\mu(G)\leqslant 8.$$

ISBN 0-201-13505-1

Consider now the 1-factors of a bipartite graph—i.e., a graph whose vertex set can be partitioned into two subsets so that every edge has one of its vertices in one set and its other vertex in the other. In the context of 1-factors, each of the subsets contains half of the vertices. The enumeration or actual construction of 1-factors (or *perfect matchings*, as they are sometimes called) of bipartite graphs has many applications—for example, in maximal flow problems and in assignment and scheduling problems arising in operational research.

Let G be a bipartite graph with $2m$ vertices, and suppose that no edge joins two vertices in $\{v_1,\dots,v_m\}$ nor two vertices in $\{v_{m+1},\dots,v_{2m}\}$. Then the adjacency matrix of G has the form

$$A(G)=\begin{bmatrix} 0 & B(G) \\ B(G)^{\mathrm{T}} & 0 \end{bmatrix},$$

where $B(G)$ is an $m\times m$ submatrix of $A(G)$. It is easy to see (Problem 13) that

$$\operatorname{haf}(A(G))=\operatorname{per}(B(G)),$$

and thus [154]

$$\begin{aligned}\mu(G)&=\operatorname{per}(B(G))\\ &=\sqrt{\operatorname{per}(A(G))}\end{aligned}$$

(cf. the condition for equality in Theorem 3 [176]). Gal and Breitbart [244] devised an algorithm for constructing 1-factors of a bipartite graph. The algorithm provides also a method for computing the permanents of (0,1)-matrices (see Section 7.3).

The use of permanents in physics was pioneered by Caianiello (see [59], [60] and the references therein) in connection with renormalization problems in quantum field theory. He used permanents and hafnians to express in algebraic form perturbation expansions of field theory for describing boson propagators, in the same way as determinants and pfaffians appear in the expansions related to fermion propagators.

The most important applications of permanents are in the areas of physics and chemistry where statistical methods are used to study phenomena that are the outcome of the combined action of a very large number of items. Such problems in order–disorder statistics, in solid-state chemistry, and in physical chemistry can be formulated as enumeration problems involving lattices. These in turn can be sometimes solved in terms of

ISBN 0-201-13505-1

permanents of corresponding incidence matrices. At this point, the pure mathematician considers the problem essentially solved, whereas the physicist or the chemist cannot see any relevance of such formulas to his problem unless and until the permanents of incidence matrices can be efficiently computed. Unfortunately, this is seldom possible, since the matrices involved are, in general, very large. In fact, in some cases—for example, in the dimer problem (*vide infra*)—the key constant to be evaluated is a limit of a function involving permanents as this order tends to infinity. In any event, a direct evaluation of the permanents by Ryser's method is out of the question. In some cases it may be possible to transform the matrix so that the determinant of the transformed matrix equals the permanent of the original matrix. If not, the problem can be solved only approximately by the use of bounds for permanents. In what follows we shall encounter both contingencies.

Problems of a structural type, such as the Ising model of ferromagnetism, the physical theory of crystals, and the adsorption of diatomic molecules on various lattices, can be expressed in terms of enumerations of dimer coverings. The dimer problem can be formulated mathematically as follows.

Consider a d-dimensional rectangular lattice graph G—i.e., a graph whose vertices are the lattice points $(x_1,\ldots,x_d)$ in a d-dimensional euclidean space whose coordinates are integers satisfying $0 \leqslant x_i \leqslant a_i$, $i = 1,\ldots,d$, where the a_i are positive integers. Let $\alpha = (a_1,\ldots,a_d)$. Two vertices of G are joined by an edge if and only if the edge is of unit length. The problem is to determine f_α, the number of 1-factors of G. A *dimer*, which represents a diatomic molecule, is a one-dimensional lattice graph consisting of two vertices and one edge of unit length. In terms of dimers f_α is the number of ways the graph G can be covered by dimers so that each dimer covers two adjacent vertices and every vertex is covered by exactly one dimer.

The problem also can be formulated as follows [142]. An *n-brick* is a d-dimensional, $d \geqslant 2$, rectangular parallelopiped of volume n with sides $a_1,\ldots,a_d$, where the a_i are integers. A *dimer* here is a 2-brick. The problem, if n is even, is to determine the number of ways of dissecting an n-brick into dimers. We first give Hammersley's [142] solution of the general case in terms of permanents. Since we do not know, except in the case $d = 2$, how to evaluate these permanents, the solution for $d \geqslant 3$ must be considered as approximate.

Partition the n-brick into n unit cubes designated by $t_1,\ldots,t_n$. Let the cubes be colored alternately black and white, and let $A = (a_{ij})$ be the incidence matrix for the configuration of the cubes: $a_{ij} = 1$ or 0 according as the cubes t_i and t_j have adjacent faces or not. All the row sums and column sums of this matrix, except those corresponding to cubes on the

ISBN 0-201-13505-1

surface, equal $2d$. Let σ be a permutation for which

$$\prod_{i=1}^{n} a_{i\sigma(i)}=1.$$

The cubes t_i and $t_{i\sigma(i)}$ are adjacent, $i=1,\dots,n$. Place a white or a black dimer on the cube t_i and the cube $t_{\sigma(i)}$ according as the cube t_i is white or black. Since there are $n/2$ white cubes, the white coloring covers all the n cubes and represents a dissection into dimers. Similarly, the black coloring gives a dissection into dimers (see Example 2.4 below). Thus, each positive diagonal of A gives rise to two dissections of the n-brick into dimers, one corresponding to the white coloring and the other to the black coloring. Conversely, any pair of dissections can be considered as white and black colorings, and thus it determines a positive diagonal of A. Hence there is a one-to-one correspondence between pairs of dissections of the n-brick and positive diagonals in A. It follows that

$$f_\alpha^2=\operatorname{per}(A). \tag{2.14}$$

Example 2.4. Consider a two-dimensional 3×4 12-brick

t_1	$\bar{t}_2$	t_3	$\bar{t}_4$
$\bar{t}_5$	t_6	$\bar{t}_7$	t_8
t_9	$\bar{t}_{10}$	t_{11}	$\bar{t}_{12}$

.

We indicate the black coloring by bars over letters. The incidence matrix of the brick is

$$A=(a_{ij})=\left[\begin{array}{cccc|cccc|cccc}
0&1&0&0&1&0&0&0&0&0&0&0\\
1&0&1&0&0&1&0&0&0&0&0&0\\
0&1&0&1&0&0&1&0&0&0&0&0\\
0&0&1&0&0&0&0&1&0&0&0&0\\
\hline
1&0&0&0&0&1&0&0&1&0&0&0\\
0&1&0&0&1&0&1&0&0&1&0&0\\
0&0&1&0&0&1&0&1&0&0&1&0\\
0&0&0&1&0&0&1&0&0&0&0&1\\
\hline
0&0&0&0&1&0&0&0&0&1&0&0\\
0&0&0&0&0&1&0&0&1&0&1&0\\
0&0&0&0&0&0&1&0&0&1&0&1\\
0&0&0&0&0&0&0&1&0&0&1&0
\end{array}\right].$$

ISBN 0-201-13505-1

We find by inspection that the diagonal corresponding to the permutation

$$\sigma=\begin{pmatrix} 1 & 2 & 3 & 4 & 5 & 6 & 7 & 8 & 9 & 10 & 11 & 12 \\ 5 & 6 & 2 & 3 & 1 & 7 & 11 & 4 & 10 & 9 & 12 & 8 \end{pmatrix}$$

is positive. The corresponding white coloring is given by

$$\begin{pmatrix} 1 & 3 & 6 & 8 & 9 & 11 \\ 5 & 2 & 7 & 4 & 10 & 12 \end{pmatrix},$$

and the corresponding dissection is

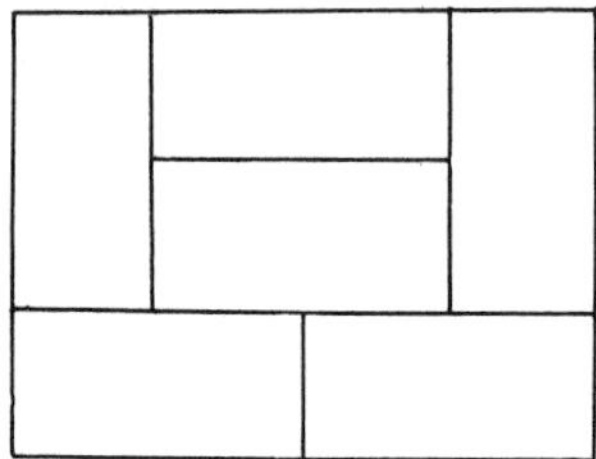

The dissection corresponding to black coloring is given by

$$\begin{pmatrix} 2 & 4 & 5 & 7 & 10 & 12 \\ 6 & 3 & 1 & 11 & 9 & 8 \end{pmatrix},$$

and thus it is

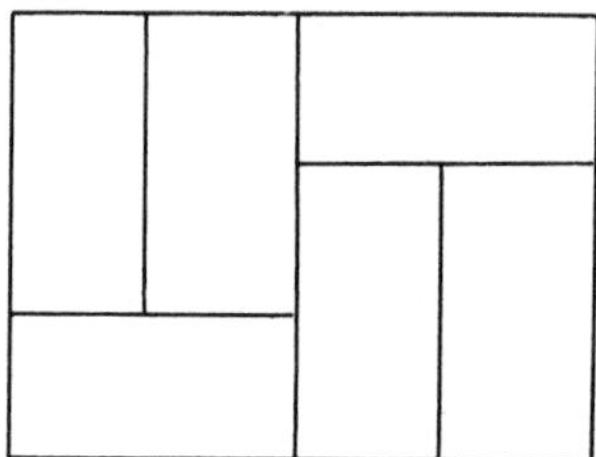

Conversely if, for example, the white and the black dissections are

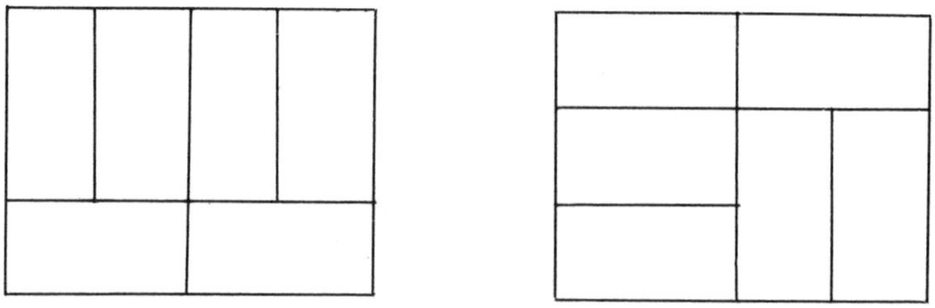

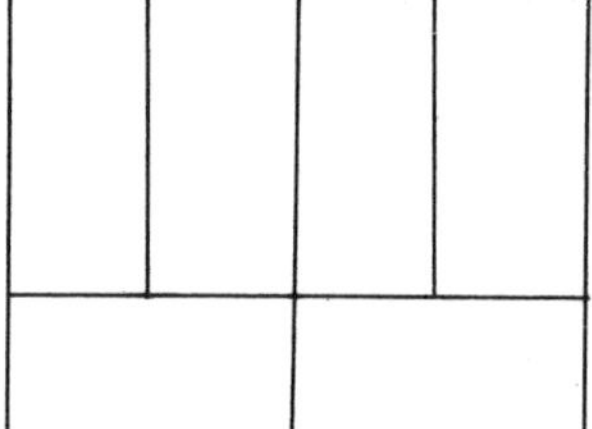

ISBN 0-201-13505-1

respectively, then we relate to it the permutation

$$\tau=\begin{pmatrix} 1 & 2 & 3 & 4 & 5 & 6 & 7 & 8 & 9 & 10 & 11 & 12 \\ 5 & 1 & 7 & 3 & 6 & 2 & 11 & 4 & 10 & 9 & 12 & 8 \end{pmatrix}$$

and we verify that the corresponding diagonal is positive.

In the general two-dimensional case of an n-brick with sides a_1 and a_2, $a_1a_2=n$, the incidence matrix (see Problem 15) is of the form

$$A=\begin{bmatrix} B & I & & & & \\ I & B & I & & 0 & \\ & I & B & I & & \\ & & \ddots & \ddots & \ddots & \\ & 0 & & I & B & I \\ & & & & I & B \end{bmatrix}, \tag{2.15}$$

where B is the $a_2\times a_2$ tridiagonal matrix

$$\begin{bmatrix} 0 & 1 & & & & \\ 1 & 0 & 1 & & 0 & \\ & 1 & 0 & 1 & & \\ & & \ddots & \ddots & \ddots & \\ & 0 & & 1 & 0 & 1 \\ & & & & 1 & 0 \end{bmatrix}$$

and I is the $a_2\times a_2$ identity matrix. Percus [165] showed that

$$\operatorname{per}(A)=\det(\tilde{A}),$$

where

$$\tilde{A}=\begin{bmatrix} B & iI & & & \\ iI & B & iI & 0 & \\ & iI & \ddots & \ddots & \\ & & \ddots & \ddots & \\ & 0 & & B & iI \\ & & & iI & B \end{bmatrix}.$$

ISBN 0-201-13505-1

Thus

$$f_{(a_1,a_2)}=\sqrt{\operatorname{per}(A)}$$
$$=\sqrt{\det(\tilde{A})}\ .$$

Therefore the number of dimer coverings of a two-dimensional lattice graph can be expressed in terms of a determinant which can be easily evaluated.

If $d \geqslant 3$ we cannot evaluate the permanent of incidence matrices, and we can only estimate the value of f_α by means of inequalities for the permanents of (0,1)-matrices.

It is known that for large values of n the function f_α grows more or less exponentially with n. In fact, Hammersley showed [142] that, if $a_i \to \infty$ for $i=1,\dots,d$, then $n^{-1}\log f_\alpha$ tends to a limit λ_d. The problem is to evaluate λ_d. It is known [142] that

$$\lambda_2=0.29156090\dots$$

and that

$$0 \leqslant \lambda_d \leqslant \tfrac{1}{2}\log d. \tag{2.16}$$

The exact value of λ_d for $d \geqslant 3$ is not known. In the all-important three-dimensional case, the right inequality in (2.16) gives

$$\lambda_3 \leqslant 0.54931. \tag{2.17}$$

Hammersley [142] obtained a lower bound for λ_3,

$$\lambda_3 \geqslant 0.418347, \tag{2.18}$$

as a particular case of the general bound,

$$\lambda_d \geqslant \frac{1}{4\pi^d}\int_0^\pi d\theta_1 \int_0^\pi d\theta_2 \cdots \int_0^\pi d\theta_d \log\left\{4\sum_{i=1}^d \sin^2\theta_i\right\}. \tag{2.19}$$

He also observed that, if the van der Waerden conjecture is valid, then

$$\lambda_d \geqslant \tfrac{1}{2}\log(2d)-\tfrac{1}{2}. \tag{2.20}$$

For $d \geqslant 4$, the bound in (2.20) is sharper than that in (2.19). Together with (2.16) it gives

$$\tfrac{1}{2}\log d - 0.153 \leqslant \lambda_d \leqslant \tfrac{1}{2}\log d. \tag{2.21}$$

ISBN 0-201-13505-1

None of the known lower bounds for the permanents for (0, 1)-matrices (see Chapter 4) produces as good a lower bound for λ_d as that in (2.18) or (2.19).

The upper bound in (2.16) was recently improved by Minc [302] by combining (2.14) with the bound in Theorem 2.1, Chapter 6. We simplify the procedure by considering, instead of an n-brick, the corresponding toroidal brick obtained by identifying and joining the opposite faces of the original brick. The incidence matrix A' of the toroidal brick has $2d$ 1's in every row and every column. Denote the corresponding value for f_α by f'_α. By (2.14) we have

$$f_\alpha \leqslant f'_\alpha = \sqrt{\operatorname{per}(A')}\ ,$$

and from Theorem 2.1, Chapter 6,

$$\operatorname{per}(A') \geqslant (2d)!^{n/2d}.$$

Hence

$$\begin{aligned}\lambda_d &= \lim_{a_i\to\infty} n^{-1}\log f_\alpha \\ &\leqslant \lim_{a_i\to\infty} n^{-1}\log(2d)!^{n/4d} \\ &= \log(2d)!^{1/4d}.\end{aligned}$$

Thus

$$\lambda_d \leqslant \tfrac{1}{2}\log(2d)!^{1/2d}. \tag{2.22}$$

For $d=3$ the inequality (2.22) yields

$$\lambda_3 \leqslant 0.548271$$

which is sharper than the bound in (2.17). In fact, the bound in (2.22) is sharper than the upper bound in (2.16) for $d \geqslant 3$, since

$$(2d)!^{1/2d} < d.$$

In applications of permanents the central difficulty is the lack of an efficient method for the evaluation of the permanents of incidence matrices. Even the known bounds for permanents, particularly the lower bounds, are not sharp enough to be of much value. However, owing to the special form of the incidence matrices which occur in the dimer problem and related problems, the evaluation of the permanents of such matrices,

ISBN 0-201-13505-1

or at least obtaining improved bounds for them, does not seem to be as intractable as it is in the general case. This is the direction in which the future research on permanents is bound to develop.

8.3 Conjectures and Unsolved Problems–Vintage 1965

In 1965, Marcus and Minc [103] listed thirteen conjectures and two unsolved problems on permanents. The list created a lot of interest, and several of the conjectures have been resolved since its publication. Here is the current status of each conjecture and each problem.

CONJECTURE 1 (*The van der Waerden Conjecture*). *If $A \in \Omega_n$, then*

$$\operatorname{per}(A) \geqslant n!/n^n,$$

where equality holds if and only if $A = J_n$.

Unresolved (see Chapter 5).

CONJECTURE 2. *If $A \in \Omega_n$, then*

$$\operatorname{per}(AA^{\mathrm{T}}) \leqslant \operatorname{per}(A).$$

The conjecture is false. Counter-example due to M. Newman: If

$$A = \frac{1}{2}\begin{bmatrix} 1 & 1 & 0 & 0 \\ 0 & 1 & 1 & 0 \\ 0 & 0 & 1 & 1 \\ 1 & 0 & 0 & 1 \end{bmatrix},$$

then $\operatorname{per}(A) = \frac{1}{8}$, whereas $\operatorname{per}(AA^{\mathrm{T}}) = \frac{9}{64} > \frac{1}{8}$.

CONJECTURE 3 (*H. J. Ryser*). *If A is an mk-square* (0,1)*-matrix with k* 1's *in each row and column, then*

$$\operatorname{per}(A) \leqslant \operatorname{per}\left(\sum_1^{m} {}^{\cdot} J\right).$$

Proved; an immediate consequence of Theorem 2.1, Chapter 6 (see Conjecture 4).

CONJECTURE 4 (*H. Minc*). *If A is an n-square* (0,1)*-matrix with row sums $r_1, \ldots, r_n$, then*

$$\operatorname{per}(A) \leqslant \prod_{i=1}^{n} (r_i!)^{1/r_i}.$$

Proved by Brégman [219] (see Theorem 2.1, Chapter 6).

ISBN 0-201-13505-1

Conjecture 5 (*H. J. Ryser*). *If the totality of v-square $(0,1)$-matrices with k 1's in each row and column contains incidence matrices of (v,k,λ)-configurations, then the permanent is minimal in this totality for one of these incidence matrices.*

Unresolved. There is some evidence to suggest that the conjecture may be false. On the one hand, in Λ_{15}^{7} there are three $(15,7,3)$-configurations whose incidence matrices have unequal permanents. On the other hand, in Λ_7^3 there exist matrices that do not represent a $(7,3,1)$-configuration and whose permanents equal that of the incidence matrix of a $(7,3,1)$-configuration.

Conjecture 6 (*H. Minc*). *Let Λ_n^k denote the class of all n-square $(0,1)$-matrices with k 1's in each row and column. Then for a fixed n,*

$$\min_{A\in\Lambda_n^k} \operatorname{per}\left(\frac{1}{k}A\right)$$

is monotone decreasing in k.

Unresolved. No progress.

Conjecture 7 (*M. Marcus and M. Newman*). *If A is n-square doubly stochastic, then*

$$\operatorname{per}(I_n-A)\geqslant 0.$$

The conjecture was proved by Brualdi and Newman [110] for any nonnegative matrix A whose spectral radius does not exceed 1. We give here a particularly simple proof due to Gibson [111].

Theorem 3.1. *If A is an $n\times n$ substochastic matrix, then*

$$\operatorname{per}(I_n-A)\geqslant 0. \tag{3.1}$$

Proof. By Ryser's formula [Section 7.2, formula (2.3)],

$$\operatorname{per}(I_n-A)=\sum_{t=0}^{n-1}(-1)^t\sum_{X\in\Lambda_{n-t}}\prod_{i=1}^{n}r_i(X),$$

where

$$\Lambda_{n-t}=\{(I_n-A)(-|\omega)\,|\,\omega\in Q_{t,n}\};$$

that is, Λ_{n-t} is the totality of $n\times(n-t)$ submatrices of I_n-A consisting of $n-t$ distinct columns of I_n-A. For $\omega\in Q_{t,n}$, let $B_\omega=(I_n-A)(-|\omega)$. The ith row sum of B_ω is nonnegative or nonpositive according as the ith

ISBN 0-201-13505-1

column of $I_n - A$ is or is not a column of the submatrix B_ω. Hence there are at least $n-t$ row sums of B_ω that are nonnegative and at least t that are nonpositive. Therefore

$$(-1)^t \prod_{i=1}^{n} r_i(B_\omega) \geq 0,$$

and thus

$$\operatorname{per}(I_n - A) \geq 0. \qquad \blacksquare$$

In [141] Gibson proved that, if A is substochastic, then

$$\operatorname{per}(I_n - A) \geq \det(I_n - A) \geq 0. \tag{3.2}$$

CONJECTURE 8 (*M. Marcus and M. Newman*). *If A is a positive semi-definite hermitian $n \times n$ matrix and $1 \leq p \leq n$, then*

$$\operatorname{per}(A) \geq \operatorname{per}(A[1,\ldots,p])\operatorname{per}(A[p+1,\ldots,n]). \tag{3.3}$$

This conjecture, the permanent analogue of the Fischer inequality for determinants, was proved by Lieb [114] as a corollary to the following remarkable result.

THEOREM 3.2 [114]. *Let $A = (a_{ij})$ be a positive semi-definite hermitian $n \times n$ matrix, and let*

$$A = \begin{bmatrix} B & C \\ C^* & D \end{bmatrix},$$

where B is $p \times p$ and D is $(n-p)\times(n-p)$. Let

$$A(\lambda) = \begin{bmatrix} \lambda B & C \\ C^* & D \end{bmatrix},$$

where λ is an indeterminate over $\mathbf{C}$. Then all the coefficients of the polynomial per(A(λ)) are real and nonnegative.

The following proof is due to Đoković [151].

Proof. Let α and β be permutations of the set $\{1,\ldots,p\}$, and let σ and τ be permutations of the set $\{p+1,\ldots,n\}$. Denote $\alpha(i)$ and $\beta(i)$ by α_i and β_i, $i=1,\ldots,p$, and $\sigma(i)$ and $\tau(i)$ by σ_{i-p} and τ_{i-p}, $i=p+1,\ldots,n$, respectively.

ISBN 0-201-13505-1

Let t be an integer, $1 \leqslant t \leqslant p$, and let

$$\alpha' = (\alpha_1, \ldots, \alpha_t), \qquad \alpha'' = (\alpha_{t+1}, \ldots, \alpha_p),$$

$$\beta' = (\beta_1, \ldots, \beta_t), \qquad \beta'' = (\beta_{t+1}, \ldots, \beta_p),$$

$$\sigma' = (\sigma_1, \ldots, \sigma_{p-t}), \qquad \sigma'' = (\sigma_{p-t+1}, \ldots, \sigma_{n-p}),$$

$$\tau' = (\tau_1, \ldots, \tau_{p-t}), \qquad \tau'' = (\tau_{p-t+1}, \ldots, \tau_{n-p}).$$

Let

$$\operatorname{per}(A(\lambda)) = \sum_{t=0}^{p} c_t \lambda^t.$$

The coefficient c_t is the sum of all diagonal products involving exactly t entries of B, $p-t$ entries of C, $p-t$ entries of C^*, and $n-2p+t$ entries of D. We have

$$mc_t = \sum_{\alpha,\beta,\sigma,\tau} a_{\alpha'\beta'} a_{\alpha''\tau'} a_{\sigma'\beta''} a_{\sigma''\tau''},$$

where $m = t!(p-t)!^2(n-2p+t)!$ and

$$a_{\alpha'\beta'} = \prod_{i=1}^{t} a_{\alpha_i\beta_i},$$

$$a_{\alpha''\tau'} = \prod_{i=1}^{p-t} a_{\alpha_{t+1}\tau_i}, \qquad \text{etc.}$$

Now A is positive semi-definite, and hence it is a Gram matrix. Let

$$x^i = (x_{i1}, \ldots, x_{in}),$$

$i = 1, \ldots, n$, be n-tuples such that

$$\begin{aligned} a_{ij} &= (x^i, x^j) \\ &= \sum_{k=1}^{n} x_{ik} \bar{x}_{jk}, \end{aligned}$$

$i,j = 1, \ldots, n$. Then

$$mc_t = \sum_{\alpha,\beta,\sigma,\tau} \sum_{I,J,H,K} X_{\alpha' I} \bar{X}_{\beta' I} X_{\alpha'' J} \bar{X}_{\tau' J} X_{\sigma' H} \bar{X}_{\beta'' H} X_{\sigma'' K} \bar{X}_{\tau'' K},$$

ISBN 0-201-13505-1

where $X_{\alpha' I}$ denotes the product $\prod_{s=1}^{t} x_{\alpha_s i_s}$, etc., and I,J,H,K denote the sequences $I=(i_1,\ldots,i_t)$, $J=(j_1,\ldots,j_{p-t})$, $H=(h_1,\ldots,h_{p-t})$, and $K=(k_1,\ldots,k_{n-2p+t})$, where each index runs through the values $1,\ldots,n$. Now, changing the order of summation, we get

$$mc_t=\sum_{I,K}\left(\sum_J\left(\sum_\alpha X_{\alpha' I}X_{\alpha'' J}\right)\left(\sum_\tau \overline{X}_{\tau' J}\overline{X}_{\tau'' K}\right)\right)\left(\sum_H\left(\sum_\beta \overline{X}_{\beta' I}\overline{X}_{\beta'' H}\right)\left(\sum_\sigma X_{\sigma' H}X_{\sigma'' K}\right)\right)$$

$$=\sum_{I,K}\left|\sum_J\left(\sum_\alpha X_{\alpha' I}X_{\alpha'' J}\right)\left(\sum_\tau \overline{X}_{\tau' J}\overline{X}_{\tau'' K}\right)\right|^2.$$

Hence c_t is real and nonnegative. ∎

Conjecture 8 is an immediate corollary of Theorem 3.2. For, setting $\lambda=1$, we have

$$\begin{aligned}\operatorname{per}(A)&=\sum_{t=0}^{p} c_t\\ &\geqslant c_p\\ &=\operatorname{per}(B)\operatorname{per}(D).\end{aligned}$$

Marcus and Soules [130] improved Lieb's result (Conjecture 8): If A is the matrix in Theorem 3.2, then

$$\operatorname{per}(A)\geqslant \operatorname{per}(B)\operatorname{per}(D)+\alpha_n^{n-2}\|C\|^2 \tag{3.4}$$

where the norm is Euclidean, and α_n denotes the least eigenvalue of A.

CONJECTURE 9 (*M. Marcus*). *Let A be a positive semi-definite hermitian $mp\times mp$ matrix partitioned as follows:*

$$A=\begin{bmatrix} A_{11} & A_{12} & \cdots & A_{1m}\\ A_{21} & A_{22} & \cdots & A_{2m}\\ \vdots & \vdots & & \\ A_{m1} & A_{m2} & \cdots & A_{mm}\end{bmatrix}, \tag{3.5}$$

in which each A_{ij} is $p\times p$. Let G be the $m\times m$ matrix whose (i,j) entry is $\operatorname{per}(A_{ij})$. *Then*

$$\operatorname{per}(A)\geqslant \operatorname{per}(G). \tag{3.6}$$

If A is positive definite, then equality holds in (3.6) *if and only if $A=\sum_{i=1}^{m} A_{ii}$.*

ISBN 0-201-13505-1

The conjecture is still unresolved for $m>2$. For $m=2$ it was proved by Lieb [114] as a corollary to his Theorem 2.2: Using the notation of Theorem 3.2, we have

$$\begin{aligned}\operatorname{per}(A) &= \sum_{t=0}^{p} c_t \\ &\geqslant c_p + c_0 \\ &= \operatorname{per}(B)\operatorname{per}(D) + \operatorname{per}(C)\operatorname{per}(C^*).\end{aligned}$$

Đoković [151] showed that for $m=2$ equality holds in (3.6) if both A_{11} and A_{22} (that is, B and D) are positive definite, even in the case when A itself is singular.

CONJECTURE 10 (*M. Marcus*). *Let* $A=(a_{ij})$ *and* $a_{ij}>0, i,j=1,\ldots,n$. *If the* $n!$ *terms* $\prod_{i=1}^{n} a_{i\sigma(i)}$ *take at most* r *different values, then* $\rho(A)\leqslant r$.

Proved by Elliott [175].

CONJECTURE 11 (*M. Marcus and M. Newman*). *There exists no doubly stochastic matrix A such that*

$$\begin{aligned}\operatorname{per}(A(i|j)) &= \operatorname{per}(A), && \text{if } a_{ij}\neq 0,\\ \operatorname{per}(A(i|j)) &\geqslant \operatorname{per}(A), && \text{if } a_{ij}=0,\end{aligned}$$

with strict inequality for at least one pair (i,j).

The above conjecture is an amended version of the original Conjecture 11 (see footnote in [128] and Theorem 2.3, Chapter 5). The conjecture is unresolved.

CONJECTURE 12 (*M. Marcus*). *The group of nonsingular generalized permutation matrices (that is, matrices of the form PD, where P is a permutation matrix and D is a diagonal matrix) is a maximal group on which the permanent is multiplicative.*

Beasley [150, 171] proved Conjecture 12 for matrices with complex entries. Beasley and Cummings [218, 289] extended the result to other fields and integral domains (with some restrictions on the characteristic and on the number of elements).

CONJECTURE 13 (*M. Marcus*). *If A is doubly stochastic and* $f(z)=\operatorname{per}(zI_n - A)$, *then the zeros of* $f(z)$ *lie in or on the boundary of the disc* $|z|\leqslant 1$.

The conjecture was proved for real zeros by Brualdi and Newman [110]. The general case was established by Brenner and Brualdi [121], who proved the following result.

ISBN 0-201-13505-1

THEOREM 3.3. *If A is a substochastic matrix, then the roots of the equation* $\operatorname{per}(zI-A)=0$ *satisfy* $0 \leqslant |z| \leqslant 1$.

The proof is based on a result of Brenner [58] stating that the permanent of a matrix with a dominant diagonal cannot vanish. Now, if A is substochastic and if $|z|>1$, then the matrix $zI-A$ has dominant main diagonal and $\operatorname{per}(zI-A) \neq 0$; that is, z cannot be a root of $\operatorname{per}(zI-A)=0$. Incidentally, the same theorem of Brenner implies Theorem 3.1 as well (Problem 19).

PROBLEM 1. *Find the maximum value of* $(U)=\operatorname{per}(U^*AU)$ *as U runs over all n-square unitary matrices. Here A is a fixed n-square positive semi-definite hermitian matrix.*

No progress beyond the Marcus-Minc [93, 102, 116] upper bound (Theorem 4.1, Chapter 6).

PROBLEM 2. *Let H be a subgroup of S_n, and let χ be a character of degree 1 of H. Under what conditions on χ and H will the inequality*

$$\sum_{\sigma \in H} \chi(\sigma) \prod_{i=1}^{n} a_{i\sigma(i)} \leqslant \operatorname{per}(A)$$

hold for all positive semi-definite hermitian A?

Unsolved.

The score: Out of fifteen conjectures and problems, seven were proved to be true, one was shown to be false, and seven remain unsolved. A new updated list of conjectures and unsolved problems on permanents is given in the next section.

8.4 Conjectures and Unsolved Problems–A Current List

It was Charles F. Kettering who said that a problem well stated is a problem half solved. For some reason this dictum does not seem to apply to permanents. Indeed, as we noted in 1965 [103], conjectures involving permanents seem to separate into two quite unsatisfactory classes; one of them contains various assertions that sound plausible but appear at present beyond reach, and the other consists in conjectures that seemed for a time undeniably true but for which counter-examples were eventually discovered. Be that as it may, we feel that it is proper to conclude this monograph with a current list of well-stated conjectures and problems, none of which has been even half solved. They are numbered so that the items surviving from the 1965 list (Section 8.3) retain their original numbering.

ISBN 0-201-13505-1

CONJECTURE 1 (*B. L. Van der Waerden*). *If A is an $n\times n$ doubly stochastic matrix, then*

$$\operatorname{per}(A)\geqslant n!/n^n, \tag{4.1}$$

with equality if and only if $A=J_n$ [103].

CONJECTURE 2. *If $A\in\Lambda_n^k, 3\leqslant k\leqslant n-1$, then*

$$\operatorname{per}(A)>n!(k/n)^n. \tag{4.2}$$

CONJECTURE 3 (*M. Marcus and H. Minc*). *If S is a doubly stochastic $n\times n$ matrix, $n\geqslant 2$, then*

$$\operatorname{per}(S)\geqslant \operatorname{per}\left(\frac{nJ_n-S}{n-1}\right). \tag{4.3}$$

If $n\geqslant 4$, equality can hold in (4.3) *if and only if $S=J_n$* [128].

CONJECTURE 4. (*E. T. H. Wang*). *If S is a doubly stochastic $n\times n$ matrix, $n\geqslant 2$, then*

$$\operatorname{per}(S)\geqslant \operatorname{per}\left(\frac{nJ_n+S}{n+1}\right). \tag{4.4}$$

If $n\geqslant 3$, equality can hold in (4.4) *if and only if $S=J_n$* [287].

CONJECTURE 5 (*H. J. Ryser*). *If the set of v-square* $(0,1)$-*matrices with k 1's in each row and column contains incidence matrices of (v,k,λ)-configurations, then the permanent is minimal in this set for one of these incidence matrices* [103].

CONJECTURE 6. (*H. Minc*). *For a fixed v,*

$$\min_{A\in\Lambda_v^k} \operatorname{per}\left(\frac{1}{k}A\right)$$

is monotone decreasing in k [103].

CONJECTURE 7 (*R. F. Scott*). *Let $x_1,\ldots,x_n$ and $y_1,\ldots,y_n$ be the distinct roots of $x^n-1=0$ and of $y^n+1=0$, respectively. Let A be the $n\times n$ matrix whose (i,j) entry is $1/(x_i-y_j), i,j=1,\ldots,n$. Then* [12],

$$|\operatorname{per}(A)| = \begin{cases} n(1\cdot 3\cdot 5\cdots(n-2))^2/2^n, & \text{if } n \text{ is odd,}\\ 0, & \text{if } n \text{ is even.}\end{cases} \tag{4.5}$$

ISBN 0-201-13505-1

CONJECTURE 8 (*P. Erdös and A. Rényi*). *If k is a fixed integer, $k \geqslant 3$, then*

$$\liminf(\operatorname{per}(A))^{1/n} > 1 \tag{4.6}$$

for A in Λ_n^k [140].

CONJECTURE 9 (*M. Marcus*). *Let A be an $mk \times mk$ positive semi-definite hermitian matrtix partitioned as follows:*

$$A = \begin{bmatrix} A_{11} & A_{12} & \cdots & A_{1m} \\ A_{21} & A_{22} & \cdots & A_{2m} \\ \vdots & & \ddots & \vdots \\ A_{m1} & A_{m2} & \cdots & A_{mm} \end{bmatrix}$$

in which each A_{ij} is k-square. Let G be the $m \times m$ matrix whose (i,j) entry is $\operatorname{per}(A_{ij})$. *Then*

$$\operatorname{per}(A) \geqslant \operatorname{per}(G). \tag{4.7}$$

If the A_{ii} are positive definite, then equality holds in (4.7) *if and only if* $A = A_{11} \dotplus \cdots \dotplus A_{mm}$ [103].

The original condition for equality included a proviso that A be positive definite. It is now known [151] that in the case $m=2$ the condition is necessary also when A_{11} and A_{22} are positive definite; the matrix itself may be singular. We altered the conjectured condition for equality accordingly.

CONJECTURE 10 (*H. Tverberg*). *If A is a doubly stochastic $n \times n$ matrix, $A \neq J_n$, then*

$$\sigma_t(A) > \sigma_t(J_n), \tag{4.8}$$

where $\sigma_t(A)$ denotes the sum of all subpermanents of A of order t [88].

CONJECTURE 11 (*M. Marcus and M. Newman*). *There exists no doubly stochastic matrix A such that*

$$\operatorname{per}(A(i|j)) = \operatorname{per}(A) \quad \textit{for all } i,j \textit{ for which } a_{ij} \neq 0,$$

$$\operatorname{per}(A(i|j)) \geqslant \operatorname{per}(A) \quad \textit{for all } i,j \textit{ for which } a_{ij} = 0,$$

ISBN 0-201-13505-1

and

$$\operatorname{per}(A(i|j)) > \operatorname{per}(A)$$

for some s,t *for which* $a_{st}=0$ [103].

CONJECTURE 12 (*D. Ž. Đoković*). *If* A *is a doubly stochastic* $n\times n$ *matrix,* $A\neq J_n$, *and* $1\leqslant k\leqslant n$, *then* [123]

$$\sigma_k(A) > \frac{(n-k+1)^2}{nk}\sigma_{k-1}(A). \tag{4.9}$$

CONJECTURE 13 (*R. Sinkhorn*). *If* A *is a doubly stochastic matrix and if* $\operatorname{per}(A(i|j))\geqslant \operatorname{per}(A)$ *for all* i,j, *then either* $A=J_n$ *or, up to permutations of rows and columns,* $A=\frac{1}{2}(I_n+P)$ [286].

CONJECTURE 14 (*R. A. Brualdi and T. H. Foregger*). *If* A *is an n-square* $(0,1)$-*matrix all of whose permanental minors of order* $n-1$ *have a common nonzero value, then* $A=J$ *or* I_n+P [260].

CONJECTURE 15 (*T. H. Foregger*). *If* A *is a nearly decomposable doubly stochastic* $n\times n$ *matrix, then*

$$\operatorname{per}(A)\geqslant 1/2^{n-1},$$

where equality holds if and only if $A=\frac{1}{2}(I_n+P)$ *up to permutations of rows and columns* [242].

CONJECTURE 16 (*H. Minc*). *If* $A\in\Lambda_n^k$, $3\leqslant k\leqslant n$, *then there exists a positive number* $\epsilon=\epsilon(k)$, *independent of n, such that*

$$\operatorname{per}(A)>(1+\epsilon)^n.$$

CONJECTURE 17 (*T. H. Foregger*). *For any n there exists an integer* $k=k(n)$ *such that*

$$\operatorname{per}(A^k)\leqslant \operatorname{per}(A)$$

for every A *in* Ω_n.

CONJECTURE 18 (*R. Merris*). *If* A *is a doubly stochastic* $n\times n$ *matrix, then* [235]

$$n\operatorname{per}(A)\geqslant \min_i \sum_{j=1}^{n} \operatorname{per}(A(j|i)). \tag{4.10}$$

ISBN 0-201-13505-1

CONJECTURE 19 (*E. T. H. Wang*). *If the permanents of two $n \times n$ Hadamard matrices are equal, then either matrix can be obtained from the other by*: (1) *permuting rows or columns*, (2) *multiplying rows or columns by* -1, *and possibly* (3) *transposing the matrix* [257].

CONJECTURE 20 (*B. Gyires*). *Let $A \in \Omega_n$, then*

$$\frac{4(\operatorname{per} A)^2}{\operatorname{per} AA^* + \operatorname{per} A^*A + 2\operatorname{per} A^2} \geqslant \frac{n!}{n^n}. \tag{4.11}$$

Equality holds in (4.11) *if and only if $A = J_n$* [276].

PROBLEM 1. *Find the maximum value of* $\operatorname{per}(U^*AU)$ *if A is a fixed n-square positive semi-definite hermitian matrix and U runs over all $n \times n$ unitary matrices* [103].

PROBLEM 2. *Let H be a subgroup S_n, and let χ be a character of degree* 1 *of H. Under what conditions on χ and H does the inequality*

$$\sum_{\sigma \in H} \chi(\sigma) \prod_{i=1}^{n} a_{i\sigma(i)} \leqslant \operatorname{per}(A)$$

hold for all positive semi-definite hermitian A? [103].

PROBLEM 3 (*B. Greenstein*). *Find all the values that* $\operatorname{per}(A)$ *can take for A in Λ_n^3.*

PROBLEM 4. *Find the maximal value of* $\operatorname{per}(A)$ *in Λ_n^k in case k does not divide n.*

PROBLEM 5 (*E. T. H. Wang*). *Can the permanent of an $n \times n$ Hadamard matrix vanish for $n > 2$?* [257].

PROBLEM 6 (*E. T. H. Wang*). *Find a significant upper bound for* $|\operatorname{per}(A)|$ *in the set of n-square* $(1, -1)$*-matrices.* [257].

PROBLEM 7 (*E. T. H. Wang*). *Do there exist, for every $n \geqslant 4$, nonsingular n-square* $(1, -1)$*-matrices A such that* $|\operatorname{per}(A)| = |\det(A)|$? [257].

PROBLEM 8 (*S. Friedland and H. Minc*). *Find matrices A on the boundary of Ω_n, other than a permutation matrix P or $(J-P)/(n-1)$, so that the permanent be monotone increasing on the segment $(1-\theta)J_n + \theta A$, $0 \leqslant \theta \leqslant 1$* [297].

PROBLEM 9. *Find a positive number $b = b(n)$ such that* $\operatorname{per}(A) > n!/n^n$ *for any $A \in \Omega_n$ satisfying $\|A - J_n\| \leqslant b$.*

PROBLEM 10. *Find numbers m and M such that*

$$2.31^n < m^n \leqslant \operatorname{per}(A) \leqslant M^n < 2.99^n \tag{4.12}$$

ISBN 0-201-13505-1

for all $A \in \Lambda_n^6$ *and sufficiently large* n. *Alternatively, find* m *and* M *that satisfy* (4.12) *for all circulants in* Λ_n^6 *for sufficiently large* n.

Inequalities (4.12) would improve known bounds for the three-dimensional dimer problem [142, 302].

Problems

1. Show that Theorem 2.3, Chapter 5, together with the inequality (1.3), Section 8.1, implies the van der Waerden conjecture for the class of doubly stochastic matrices for which the inequality (1.3) holds.
2. Show that if, $A \in \Omega_n$, then

$$\sigma_2(A) = \tfrac{1}{2}\left(n(n-2) + \|A\|^2\right)$$

where $\|A\|^2 = \sum_{i,j} a_{ij}^2$. (*Hint*: Use the technique in the proof of Theorem 4.1, Chapter 5.)

3. Deduce from the result in Problem 2 that, if $A \in \Omega_n$ and $A \neq J_n$, then

$$\sigma_2(A) > \sigma_2(J_n).$$

4. Let A be an $n \times n$ (0,1)-matrix all of whose subpermanents of order $n-1$ have a common nonzero value. Show that A is fully indecomposable and that $A \in \Lambda_n^r$—that is, that A is a nonzero multiple of a fully indecomposable, doubly stochastic matrix.
5. Prove the inequality (1.13) in the case $\alpha \in Q_{k,n}$.
6. Prove Hall's inequality

$$L(r,n) \geqslant \prod_{j=1}^{r} (n+1-j)!.$$

7. Find upper bounds for $L(6,6)$, $L(7,7)$, and $L(8,8)$: (a) using formula (2.2); (b) using formula (2.4), together with Theorem 2.1, Chapter 6. Compare the obtained bounds with the actual values [259]:

$$L(6,6) = 6!5!(9408); \quad L(7,7) = 7!6!(16942080);$$
$$L(8,8) = 8!7!(535281401856).$$

8. Find a lower bound for $L(8,8)$ using formulas (2.7) and (2.8).
9. Find lower bounds for $L(6,6)$, $L(7,7)$, and $L(8,8)$ using formulas (2.5) and (2.6). Compare the bounds found and the bound obtained in Problem 8 with the values given in Problem 7.
10. Let G be a complete graph with six vertices. Use the inequality (2.12) to compute an upper bound for $\mu(G)$, the number of 1-factors of G.
11. Let G be the graph in Problem 10. Use the inequality (2.13) to obtain an upper bound for $\mu(G)$.

ISBN 0-201-13505-1

12. Let G be the graph in Problem 10. Evaluate the hafnian of the matrix $J-I_6$ by constructing all the 1-factors of G.
13. Let

$$A=\begin{bmatrix} 0 & B \\ B^{\mathrm{T}} & 0 \end{bmatrix},$$

where B is square. Show that $\operatorname{haf}(A)=\operatorname{per}(B)$.
14. Let G be the 12-brick in Example 2.4, and let φ be the permutation

$$\begin{pmatrix} 1 & 2 & 3 & 4 & 5 & 6 & 7 & 8 & 9 & 10 & 11 & 12 \\ 2 & 1 & 7 & 3 & 6 & 10 & 8 & 4 & 5 & 9 & 12 & 11 \end{pmatrix}.$$

Verify that the diagonal of the incidence matrix of G that corresponds to φ is positive. Find the pair of dissections of G into dimers determined by φ.
15. Let G be a two-dimensional n-brick with sides of lengths a and b, $n=ab$. Show that the incidence matrix of G has the form shown in (2.15).
16. Let A_n be the $n\times n$ (0,1)-matrix with 1's in the (i,j) position for $i=j+1$, $i=j$, and $j=i+k$, for a fixed integer k, and zeros elsewhere. Show that [300]:

$$\operatorname{per}(A_n)=\operatorname{per}(A_{n-1})+\operatorname{per}(A_{n-k-1}).$$

17. Let A_n be the matrix defined in Problem 16. Show that; for sufficiently large n,

$$\operatorname{per}(A_n)>c^n$$

for any c, $1<c<d$, where d is the largest root of $x^{k+1}-x^k-1=0$.
18. Show that for every c, $1<c<h$, where h is the largest root of $x^t-x^{t-1}-1=0$,

$$\operatorname{per}\left(I_n+P+P^t\right)>c^n$$

for sufficiently large n.
19. Brenner [58] showed that the permanent of a matrix with a dominant diagonal cannot vanish. Prove that this result implies Theorem 3.1.
20. Show that Conjecture 4, Section 8.4, implies the van der Waerden conjecture.

ISBN 0-201-13505-1

Bibliography

1812

1. J. P. M. Binet, Mémoire sur un système de formules analytiques, et leur application à des considérations géometriques, *J. Ec. Polyt.* **9** (1812); *Cah.* **16**, 280–302.

The paper contains identities for permanents of $m \times n$ matrices for $m=2, 3, 4$. No proofs, nor general formulas.

2. A. L. Cauchy, Mémoire sur les fonctions qui ne peuvent obtenir que deux valeurs égales et de signes contraires par suite des transpositions operées entre les variables qu'elles renferment, *J. Ec. Polyt.* **10** (1812); *Cah.* **17**, 29–112, *Oeuvres* (2)i.

An important work on determinants. Introduces permanents as a class of "fonctions symetriques permanentes."

1843

3. G. Boole, On the transformation of multiple integrals, *Cambridge Math. J.* **4** (1843), 20–28.

Defines determinant of order n by defining the permanent of the matrix (without using the words "determinant" or "permanent") and then altering the signs of certain terms.

1855

4. C. W. Borchardt, Bestimmung der symmetrischen Verbindungen vermittelst ihrer erzeugenden Funktion, *Monatsb. Akad. Wiss. Berlin* (1855), 165–171; or *Crelle's J.* **53** (1855), 193–198; or *Gesammelte Werke*, 97–105.

Let $A=(a_{ij})=\left(\frac{1}{t_i-s_j}\right)$ and $B=(a_{ij}^2)=\left(\frac{1}{(t_i-s_j)^2}\right)$. Then

$$\operatorname{per}(A)\det(A)=\det(B).$$

ENCYCLOPEDIA OF MATHEMATICS and Its Applications, Gian-Carlo Rota (ed.). Vol. 6: Henryk Minc, Permanents

ISBN 0-201-13505-1

1856

5. F. Joachimstal, De aequationibus quarti et sexti gradus quae in theoria linearum et superficierum secundi gradus occurunt, *Crelle's J.* **53** (1856), 149–172.
Shows, without definite proof, how to generate Binet's identities for an $m\times n$ matrix from those for an $(m-1)\times n$ matrix. Describes permanent by stating that it "tantum a determinante differt, quod ejus termini sunt positivi."

1859

6. A. Cayley, Note sur les normales d'une conique, *Crelle's J.* **56** (1859), 182–185; or *Collected Mat. Papers* **IV**, 74–77.
Let $A=(a_{ij})$ be a 3×3 matrix, $a_{ij}\neq0$, and let B and C be 3×3 matrices whose (i,j) entries are a_{ij}^2 and a_{ij}^{-1}, respectively. Then

$$\operatorname{per}(A)\det(A)=\det(B)+2\Big(\prod_{i,j}a_{ij}\Big)\det(C).$$

1865

7. J. Horner, Notes on determinants, *Quart. J. Math.* **8** (1865), 157–162.
Let A and B be 3×3 matrices. Then

$$\det(A)\det(B)=\sum_i\det(P_i)\operatorname{per}(A*(P_iB)),$$

where the summation is over all 3×3 permutation matrices P_i, and $*$ denotes the Hadamard product. Permanents are called "conterminants."

1872

8. W. Spottiswoode, On determinants of alternate numbers, *Proc. London Math. Soc.* **7** (1872), 100–112.
Relations between determinants and permanents of 2×2 and 3×3 matrices whose entries are "alternate numbers" (that is, satisfying $ab=-ba$).

1876

9. F. Faà Di Bruno, *Théorie des formes binaires*, Turin, 1876.
Contains a verification of Borchardt's equality by Cayley.

1879

10. J. Hammond, Question 6001, *Educ. Times* **32** (1879), 179.
A proposition that the coefficients of the polynomial $\prod_{i=1}^n\sum_{j=1}^n a_{ij}x_j$ are expressible as "alternate determinants" (that is, permanents).

ISBN 0-201-13505-1

11. R. F. Scott, Notes on determinants, *Messenger of Math.* **8** (1879), 182–187.
Let A and $A^{(n)}$ be $n\times n$ matrices whose (i,j) entries are s_i-t_j and $(s_i-t_j)^n$, respectively. Then

$$\det(A^{(n)})=\frac{1}{n!}\prod_{r=1}^{n}\binom{n}{r}\prod_{i<j}(s_i-s_j)\prod_{i<j}(t_i-t_j)\operatorname{per}(A).$$

1881

12. R. F. Scott, Mathematical notes, *Messenger of Math.* **10** (1881), 142–149.
Let $A=((x_i-y_j)^{-1})$, where $x_1,\ldots,x_n$ and $y_1,\ldots,y_n$ are the distinct roots of $x^n-1=0$ and of $y^n+1=0$, respectively. The author gives without proof the identity

$$|\operatorname{per}(A)|=\begin{cases}\dfrac{n(1\cdot3\cdot5\cdots(n-2))^2}{2^n}, & \text{if } n \text{ is odd,}\\ 0, & \text{if } n \text{ is even.}\end{cases}$$

13. T. Muir, *A Treatise on Determinants*, London, 1881.
See the revision of W. H. Metzler [39].

1882

14. T. Muir, On a class of permanent symmetric functions, *Proc. Roy. Soc. Edinburgh* **11** (1882), 409–418.
Introduces the term "permanent" for square matrices and establishes properties resembling properties of determinants. Also, shows *inter alia* that the determinant of a product of two $n\times n$ matrices is expressible as an aggregate of $n!$ permanents of the same order, and that a product of the permanent of an $n\times n$ matrix and the determinant of an $n\times n$ matrix is expressible as an aggregate of $n!$ determinants of the same order.

15. V. Řehořovsky, O vytvornjíci funkci Borchardt-ove, *Časopis pro pěstování math. a fys.* **11** (1882), 111–120.
An examination of Borchardt's identity and of the associated generating function.

1890

16. T. Muir, *The Theory of Determinants in the Historical Order of Development*, Vol. I, Part I, London, 1890.
Abstracts of writings on the theory of determinants up to, but excluding, Cayley.

1893

17. M. Lerch, Krátký důkaz Borchardtovy věty determinantni, *Časopis pro pěstování math. a fys.* **23** (1893), 76–78.
Another examination of Borchardt's identity.

ISBN 0-201-13505-1

1897

18. T. Muir, A relation between permanents and determinants, *Proc. Roy. Soc. Edinburgh* **22** (1897), 134–136.
Let A be an $n\times n$ matrix. Then

$$\sum_{r=0}^{n}\sum_{\omega\in Q_{r,n}}(-1)^r\operatorname{per}(A[\omega|\omega])\det(A(\omega|\omega))=0,$$

where $\operatorname{per}(A[\omega|\omega])=1$ and $\det(A(\omega|\omega))=\det(A)$ if $r=0$, and $\operatorname{per}(A[\omega|\omega])=\operatorname{per}(A)$ and $\det(A(\omega|\omega))=1$ if $r=n$.

1899

19. F. Ferber, Sur un symbole analogue aux déterminants, *Bull. Soc. Math. France* **27** (1899), 285–288.
Permanents of Vandermonde matrices.

20. T. Muir, On a development of a determinant of the *mn*th order, *Trans. Roy. Soc. Edinburgh* **39** (1899), 623–628.
Expression of the determinant of an mn-square matrix A as a sum of permanents of $m\times m$ matrices whose entries are subdeterminants of A.

1900

21. F. Ferber, Application du symbole des déterminants positifs, *Bull. Soc. Math. France* **28** (1900), 128–130.
Two examples of the use of permanents.

22. T. Muir, On Jacobi's expansion for the difference-product when the number of elements is even, *Proc. Roy. Soc. Edinburgh* **23** (1900), 133–141.
Proves Jacobi's expansion for the difference-product (*Werke* **III**, 439–452) using pfaffians and permanents.

1903

23. A. Young, The expansion of the *n*th power of a determinant, *Messenger of Math.* **33** (1903), 113–116.
An expression for the typical term of $(\operatorname{per} A)^n$, where A is a 3-square matrix.

24. R. F. Muirhead, Some methods applicable to identities and inequalities of symmetric algebraic functions of n letters, *Proc. Edinburgh Math. Soc.* **21** (1903), 144–157.
Let $A=(a_i^{h_j})$ and $B=(a_i^{k_j})$ be $n\times n$ matrices, where the a_i are positive numbers and the h_j and k_j are nonnegative integers. Then $\operatorname{per}(A)\leqslant\operatorname{per}(B)$ if and only if the sequence $(h_1,\ldots,h_n)$ is majorized by $(k_1,\ldots,k_n)$.

ISBN 0-201-13505-1

1906

25. T. Muir, *The Theory of Determinants in the Historical Order of Development*, Vol. I (2nd ed.), London, 1906.
Includes Part I [16] published in 1890.

1911

26. T. Muir, *The Theory of Determinants in the Historical Order of Development*, Vol. II, London, 1911.
Contains a section on permanents, pp. 445–448.

1912

27. T. Muir, Note on double alternants, *Trans. Roy. Soc. S. Africa* **3** (1912), 177–185.
A relation between a determinant and a permanent.

28. T. Muir, Question 6001, *Educ. Times* **65** (1912), 139; or *Math. from Educ. Times* (2) **22** (1912), 49–50.
Expression of the coefficients in the expansion of $\prod_{i=1}^{n}\sum_{j=1}^{n}a_{ij}x_j$ by means of permanents.

1913

29. G. Pólya, Aufgabe 424, *Arch. Math. Phys.* (3) **20** (1913), 271.
There is no uniform way of affixing + and − signs to the elements of an $n\times n$ matrix, $n\geqslant 3$, so as to change the determinant into the permanent.

1915

30. T. Muir, Question 18015, *Educ. Times* **68** (1915), 238; *Math. from Educ. Times* (2) **29** (1915), 58–60; *Math. Quest. and Sol.* **1** (1915), 16.
An inequality between a permanent and a determinant.

31. T. Muir, Determinants whose elements are alternating numbers, *Messenger of Math.* **45** (1915), 21–27.
A follow-up on Spottiswoode's paper [8].

1916

32. D. König, Über Graphen und ihre Anwendung auf Determinantentheorie und Mengenlehre, *Math. Ann.* **77** (1916), 453–465; or *Math. és Termesz. Értesitö* **34** (1916), 104–119.
The permanent of a doubly stochastic matrix is positive. The permanent of a (0,1)-matrix with k 1's in each row and each column is at least k.

ISBN 0-201-13505-1

1917

33. G. Frobenius, Über zerlegbare Determinanten, *Sber. Preuss. Akad. Wiss.* (1917), 274–277.

The permanent of a nonnegative $n \times n$ matrix vanishes if and only if the matrix contains a $k \times (n-k+1)$ zero submatrix. The permanent of a doubly stochastic matrix is positive.

1918

34. I. Schur, Über endliche Gruppen und Hermitesche Formen, *Math. Z.* **1** (1918), 184–207.

The determinant of a positive semi-definite hermitian matrix cannot exceed its permanent.

1920

35. T. Muir, *The Theory of Determinants in the Historical Order of Development*, Vol. III, London, 1920.

Contains abstracts of papers [7], [8], [10], [11], and of book [9].

1923

36. T. Muir, *The Theory of Determinants in the Historical Order of Development*, Vol. IV, London, 1923.

Contains abstracts of papers [12], [14], [15], [17], [18], [19], [20].

1926

37. B. L. van der Waerden, Aufgabe 45, *Jber. Deutsch. Math. Verein.* **35** (1926), 117.

The "van der Waerden conjecture": to determine the minimum for the permanent function on the set of $n \times n$ doubly stochastic matrices.

38. H. W. Turnbull, *The Theory of Determinants, Matrices, and Invariants*, London, 1926.

Contains the definition of the permanent function and a formula involving permanents.

1928

39. T. Muir (revised by William H. Metzler), *A Treatise on the Theory of Determinants*, London, 1928.

Revised and enlarged edition of [13]. Contains several formulas involving determinants and permanents. Introduces permanents of three-way matrices.

1930

40. T. Muir, *Contributions to the History of Determinants 1900–1920*, London, 1930.

Contains abstracts of [21], [22], [23], [27], [28], [29], [30], [31], [32].

ISBN 0-201-13505-1

1931

41. E. Egerváry, Matrixok kombinatórius tulajdonságairól, *Mat. Fiz. Lapok* **38** (1931), 16–28.
Contains the König–Egerváry theorem: The minimal number of lines in a (0, 1)-matrix that contain all the 1's in the matrix is equal to the term rank of the matrix.

42. D. König, Graphok es matrixok, *Mat. Fiz. Lapok* **38** (1931), 116–119.
A generalization of the Frobenius–König theorem. The König–Egerváry theorem.

1933

43. D. König, Über trennende Knotenpunkte in Graphen (nebst Anwendungen auf Determinanten und Matrizen), *Acta Litt. ac Sci. (Sec. Sci. Math.), Szeged* **6** (1933), 155–179.
Continuation of [42] (see König's book [48]).

1934

44. G. H. Hardy, J. E. Littlewood, and G. Pólya, *Inequalities*, London, 1934.
Contains a generalization of Muirhead's theorem [24] to all nonnegative real exponents, and proofs of the generalized theorem.

45. J. Touchard, Sur un problème de permutations, *C. R. Acad. Sci. Paris* **198** (1934), 631–633.
An explicit formula for the "ménage numbers"—that is, for $\text{per}(nJ_n - I_n - P)$.

46. J. H. M. Wedderburn, *Lectures on Matrices*, Providence, 1934.
Defines permanents, develops their multilinear properties, and introduces the concept of induced matrix (pp. 75–76).

1935

47. P. Hall, On representatives of subsets, *J. London Math. Soc.* **10** (1935), 26–30.
A necessary and sufficient condition for a configuration of subsets to have an SDR; or equivalently, if the subsets are finite, for a nonnegative matrix to have a positive permanent.

1936

48. Denes König, *Theorie der endlichen und unendlichen Graphen*, Leipzig, 1936.
The section on applications of graphs to matrices and determinants (Chapter XIV, s. 3) deals really with permanents. In particular, it contains the Frobenius–König theorem, the König–Egerváry theorem, and a statement of the van der Waerden conjecture.

1939

49. A. C. Aitken, *Determinants and Matrices*, Edinburgh, 1939.
Definition and expansions of permanents. "Problème des rencontres"—that is, the evaluation of $\text{per}(nJ_n - I_n)$.

ISBN 0-201-13505-1

50. I. Kaplansky, Solution of the "problème des ménages," *Bull. Amer. Math. Soc.* **49** (1943), 784–785.

A solution of the problème des ménages—i.e., the evaluation of $2n!\ \mathrm{per}(nJ_n - I_n - P)$.

1945

51. T. Venkatarayudu, Immanents of a matrix associated with a finite abelian group, *Proc. Indian Acad. Sci., Sect. A* **21** (1945), 103–104.

All immanents of a certain matrix associated with a finite abelian group vanish, except the permanent.

1948

52. Marshall Hall, Jr., Distinct representatives of subsets, *Bull. Amer. Math. Soc.* **54** (1948), 922–926.

If A is an $m \times n$ (0,1)-matrix with nonzero permanent and row sums at least k, $k \leqslant m$, then $\mathrm{Per}(A) \geqslant k!$.

1950

53. Paul R. Halmos and Herbert E. Vaughan, The marriage problem, *Amer. J. Math.* **72** (1950), 214–215.

Contains a simple combinatorial proof of P. Hall's theorem [47].

1953

54. E. R. Caianiello, On quantum field theory, I. Explicit solution of Dyson's equation in electrodynamics without use of Feynman graphs, *Nuovo Cimento* (9) **10** (1953), 1634–1652.

Introduces hafnians and uses them in an algorithm.

55. H. B. Mann and H. J. Ryser, Systems of distinct representatives, *Amer. Math. Monthly* **60** (1953), 397–401.

A short proof of M. Hall's theorem [52], based on the method in [53].

1954

56. Pradillo Julio Garcia, On properties and applications of permanents (Spanish), *Gaceta Mat.* (1) **6** (1954), 8–14.

Includes an application to computing a formula for elementary symmetric functions.

ISBN 0-201-13505-1

1956

57. E. R. Caianiello, Proprietà pfaffiani e hafniani, *Ricerca, Napoli* **7** (1956), 25–31.
An expansion theorem for hafnians involving permanents.

1959

58. J. L. Brenner, Relations among the minors of a matrix with dominant principal diagonal, *Duke Math. J.* **26** (1959), 563–568.
A matrix with a dominant diagonal has a nonzero permanent.

59. E. R. Caianiello, Regularization and renormalization, I, *Nuovo Cimento* (10) **13** (1959), 637–661.
Uses hafnians and permanents to express expectation values of products of free commuting fields. Contains an expansion formula for permanents.

60. E. R. Caianiello, Theory of coupled fields, *Nuovo Cimento Suppl.* **14** (1959), 177–191.
Uses permanents and hafnians in perturbative expansions.

61. Jack Levine, Note on an identity of Cayley, *Amer. Math. Monthly* **66** (1959), 290–292.
Extends Cayley's identity [6] to 4×4 matrices of rank less than 3.

62. Marvin Marcus and Morris Newman, On the minimum of the permanent of a doubly stochastic matrix, *Duke Math. J.* **26** (1959), 61–72.
Important partial results on the van der Waerden conjecture. Main result: If A is a doubly stochastic positive $n \times n$ matrix with minimal permanent, then $A = J_n$.

1960

63. L. Carlitz and Jack Levine, An identity of Cayley, *Amer. Math. Monthly* **67** (1960), 571–573.
A generalization of Cayley's theorem [6]: If $A = (a_{ij})$ is an $n \times n$ matrix of rank less than 3, and if $B = (a_{ij}^{-1})$ and $C = (a_{ij}^{-2})$, then $\text{per}(B)\det(B) = \det(C)$.

64. Marvin Marcus, Some properties and applications of doubly stochastic matrices, *Amer. Math. Monthly* **67** (1960), 215–221.
Contains a short section on the van der Waerden conjecture.

65. Marvin Marcus and Morris Newman, Permanents of doubly stochastic matrices, *Proc. Symp. Appl. Math., Amer. Math. Soc.* **10** (1960), 169–174.
A brief account of results obtained in [62].

66. Paul J. Nikolai, Permanents of incidence matrices, *Math. Comput.* **14** (1960), 262–266.
Exhibits two incidence matrices of nonisomorphic (15,7,3)-designs with unequal permanents.

67. H. J. Ryser, Compound and induced matrices in combinatorial analysis, *Proc. Symp. Appl. Math., Amer. Math. Soc.* **10** (1960), 149–167.

ISBN 0-201-13505-1

The concluding page contains a short survey of problems on permanents of nonnegative matrices.

68. H. J. Ryser, Matrices of zeros and ones, *Bull. Amer. Math. Soc.* **66** (1960), 442–464.
Conjecture: In the class of vk-square $(0,1)$-matrices with row sums and column sums equal to k, the permanent function takes its maximum on the direct sum of k-square matrices of 1's.

69. M. F. Tinsley, Permanents of cyclic matrices, *Pacific J. Math.* **10** (1960), 1067–1082.
If A is a $(0,1)$-matrix which is a sum of s commuting permutation matrices, $s \geqslant 3$, and if $|\det(A)| = \operatorname{per}(A)$, then A is equal, up to permutations of rows and columns, to a direct sum of 7×7 circulants $I_7 + P + P^3$.

1961

70. Marvin Marcus and Henryk Minc, On the relation between the determinant and the permanent, *Illinois J. Math.* **5** (1961), 376–381.
There exists no linear transformation on $n \times n$ matrices, $n \geqslant 3$, that converts the determinant of each matrix into its permanent. This generalizes a result of Pólya [29].

71. Marvin Marcus, Henryk Minc, and Benjamin Moyls, Some results on nonnegative matrices, *J. Res. Nat. Bur. Standards* **65B** (1961), 205–209.
Lower bounds for the permanents of doubly stochastic matrices in terms of their Birkhoff numbers.

72. Marvin Marcus and Morris Newman, The permanent function as inner product, *Bull. Amer. Math. Soc.* **67** (1961), 223–224.
Announces the results in [79].

73. N. S. Mendelsohn, Permutations with confined displacements, *Canad. Math. Bull.* **4** (1961), 29–38.
Formulas for the permanents of $(0,1)$-circulants $P + P^2 + \cdots + P^k$, for $k = 1,2,3,4, n-2, n-1, n$. The formulas for $k = 3,4,n-2$ are incorrect (see [94]).

1962

74. Marvin Marcus, An inequality connecting the P-condition number and the determinant, *Numer. Math.* **4** (1962), 350–353.
The inequality yields a lower bound for the determinant of a positive definite hermitian matrix in terms of the product of its principal subpermanents.

75. Marvin Marcus, Linear operations on matrices, *Amer. Math. Monthly* **69** (1962), 837–847.
Contains a section on permanents: statements of results in [29], [70], and [76].

76. Marvin Marcus and F. C. May, The permanent function, *Canad. J. Math.* **14** (1962), 177–189.
Determines the form of linear transformations on $n \times n$ matrices that hold the permanent of each matrix fixed.

ISBN 0-201-13505-1

77. Marvin Marcus and Henryk Minc, Some results on doubly stochastic matrices, *Proc. Amer. Math. Soc.* **76** (1962), 571–579.

If A is a doubly stochastic $n \times n$ matrix, then $\operatorname{per}(A) \geqslant 1/n^n$.

78. Marvin Marcus and Henryk Minc, The Pythagorean theorem in certain symmetry classes of tensors, *Trans. Amer. Math. Soc.* **104** (1962), 510–515.

If $A = (a_{ij})$ is a positive semi-definite hermitian $n \times n$ matrix, then

$$\operatorname{per}(A) \geqslant \frac{n!}{n^{2n}} \prod_{i=1}^{n} a_{ii}.$$

79. Marvin Marcus and Morris Newman, Inequalities for the permanent function, *Ann. Math.* **675** (1962), 47–62.

Permanent is exhibited as an inner product in a suitable symmetry class of tensors. The van der Waerden conjecture is shown to hold for positive semi-definite doubly stochastic matrices.

80. B. N. Moyls, Marvin Marcus, and Henryk Minc, Permanent preservers on the space of doubly stochastic matrices, *Canad. J. Math.* **14** (1962), 190–194.

Determines the form of mappings described in the title.

1963

81. P. Erdös and A. Rényi, On random matrices, *Magyar Tud. Akad. Mat. Kutató Int. Közl.* **8** (1963), 455–461.

On the probability that a (0,1)-matrix with a given number of 1's will have a positive permanent.

82. Marvin Marcus, The permanent analogue of the Hadamard determinant theorem, *Bull. Amer. Math. Soc.* **69** (1963), 494–496.

If $A = (a_{ij})$ is a positive semi-definite hermitian $n \times n$ matrix, then

$$n a_{ii} \operatorname{per}(A(i|i)) \geqslant \operatorname{per}(A) \geqslant a_{ii} \operatorname{per}(A(i|i)),$$

$i = 1, \ldots, n$, and hence

$$\operatorname{per}(A) \geqslant \prod_{i=1}^{n} a_{ii}.$$

83. Henryk Minc, A note on an inequality of M. Marcus and M. Newman, *Proc. Amer. Math. Soc.* **14** (1963), 890–892.

A slight generalization and a new proof for a result in [79] (the van der Waerden conjecture for positive semi-definite symmetric doubly stochastic matrices).

84. Henryk Minc, Upper bounds for permanents of (0,1)-matrices, *Bull. Amer. Math. Soc.* **69** (1963), 789–791.

If A is a (0,1)-matrix with row sums $r_1, \ldots, r_n$, then

$$\operatorname{per}(A) \leqslant \prod_{i=1}^{n} \frac{r_i + 1}{2}.$$

Conjecture:

$$\operatorname{per}(A) \leqslant \prod_{i=1}^{n} (r_i!)^{1/r_i}.$$

ISBN 0-201-13505-1

85. L. Mirsky, Results and problems in the theory of doubly stochastic matrices, *Z. Wahrschein.* **1** (1963), 319–334.

Contains a brief survey of results on permanents.

86. D. E. Rutherford, Inverses of Boolean matrices, *Proc. Glasgow Math. Assoc.* **6** (1963), 49–53.

Introduces the concept of Boolean permanent (under the name "Boolean determinant").

87. Hebert John Ryser, *Combinatorial Mathematics*, Math. Assoc. Amer., 1963.

Contains Ryser's formula for the evaluation of permanents. Also, material on "problème des rencontres," "problème des ménages," the van der Waerden conjecture, permanents of (0,1)-matrices, etc.

88. Helge Tverberg, On the permanent of a bistochastic matrix, *Math. Scand.* **12** (1963), 25–35.

Let $\sigma_t(A)$ denote the sum of the subpermanents of A of order t. If $A \neq J_n$ is a doubly stochastic matrix, then $\sigma_t(A) \geqslant \sigma_t(J_n)$ for $t=2$ and 3.

1964

89. Marvin Marcus, The Hadamard theorem for permanents, *Proc. Amer. Math. Soc.* **15** (1964), 967–973.

Proves previously announced result [82]: If $A=(a_{ij})$ is a positive semi-definite hermitian $n\times n$ matrix, then

$$\operatorname{per}(A) \geqslant \prod_{i=1}^{n} a_{ii}.$$

90. Marvin Marcus, The use of multilinear algebra for proving matrix inequalities, *Recent Advances in Matrix Theory*, Wisconsin, 1964, pp. 61–80.

A demonstration of the use of multilinear algebra for proving some previously obtained inequalities.

91. Marvin Marcus and W. R. Gordon, Inequalities for subpermanents, *Illinois J. Math.* **8** (1964), 607–614.

Upper bounds for the sum of the squares of the absolute values of all subpermanents of order k of a matrix.

92. Marvin Marcus and Henryk Minc, *A Survey of Matrix Theory and Matrix Inequalities*, Boston, 1964.

Contains material on elementary properties of permanents and of induced matrices, various inequalities involving permanents, a list of partial results on the van der Waerden conjecture, etc.

93. Marvin Marcus and Henryk Minc, Inequalities for general matrix functions, *Bull. Amer. Math. Soc.* **70** (1964), 308–313.

If N is a normal $n\times n$ matrix with eigenvalues $r_1,\dots,r_n$, then

$$|\operatorname{per}(N)| \leqslant \frac{1}{n}\sum_{i=1}^{n} |r_i|^n.$$

Other inequalities for permanents, including a generalization of the van der Waerden conjecture to the set of $n\times n$ positive semi-definite hermitian matrices.

ISBN 0-201-13505-1

94. Henryk Minc, Permanents of (0, 1)-circulants, *Canad. Math. Bull.* **7** (1964), 253–263.
Formulas for the permanents of (0, 1)-circulants I_n+P+P^2 and $I_n+P+P^2+P^3$.

95. Hazel Perfect, An inequality for the permanent function, *Proc. Cambridge Philos. Soc.* **60** (1964), 1030–1031.
If A is a substochastic matrix, then

$$\operatorname{per}\left(\tfrac{1}{2}(I+A)\right) \leqslant \tfrac{1}{2}(1+\operatorname{per}(A)).$$

96. M. L. Slater and R. J. Thompson, A permanent inequality for positive functions on the unit square, *Pacific J. Math.* **14** (1964), 1069–1078.
The inequality referred to in the title is applied to prove the inequality for the maximal diagonal product in a doubly stochastic matrix obtained in [77].

1965

97. R. A. Brualdi and M. Newman, Inequalities for permanents and permanental minors, *Proc. Cambridge Philos. Soc.* **61** (1965), 741–746.
If $A=(a_{ij})$ is a doubly stochastic $n\times n$ matrix and $0\leqslant\alpha\leqslant 1$, then

$$\operatorname{per}(\alpha I+(1-\alpha)A)\leqslant \alpha+(1-\alpha)\operatorname{per}(A),$$
$$\sum_i (1-a_{ii})\operatorname{per}(A(i|i))\leqslant 1-\operatorname{per}(A),$$
$$\sum_{i,j}\operatorname{per}(A(i|j))\leqslant n, \text{ etc.}$$

98. R. A. Brualdi and M. Newman, Some theorems on the permanent, *J. Res. Nat. Bur. Standards* **69B** (1965), 159–163.
For a positive integer k, let $f(k)$ be the smallest order of (0, 1)-matrix with permanent equal to k. Then $\log f(k) \sim \log\log k$. The paper contains also results on permanents of nonnegative circulants.

99. C. J. Everett, An inequality on doubly stochastic matrices, *Proc. Amer. Math. Soc.* **16** (1965), 310–313.
The van der Waerden conjecture holds for doubly stochastic $n\times n$ matrices with $n-1$ identical rows.

100. Marvin Marcus, Matrix applications of a quadratic identity for decomposable symmetrized tensors, *Bull. Amer. Math. Soc.* **71** (1965), 360–364.
A lower bound for $\operatorname{per}(K[\omega|\omega])$, where K is positive definite hermitian and $\omega\in G_{m,n}$, in terms of the eigenvalues of K.

101. Marvin Marcus and Henryk Minc, Diagonal products in doubly stochastic matrices, *Quart. J. Math. Oxford Ser.* (2) **16** (1965), 32–34.
If A is an $n\times n$ doubly stochastic matrix with h eigenvalues of modulus 1, then

$$\operatorname{per}(A)\geqslant(n-h+1)^{-(n-h+1)}.$$

If A happens to be irreducible, then $\operatorname{per}(A)\geqslant(h/n)^n$.

ISBN 0-201-13505-1

102. Marvin Marcus and Henryk Minc, Generalized matrix functions, *Trans. Amer. Math. Soc.* **116** (1965), 316–329.

Proofs of results announced in [93].

103. Marvin Marcus and Henryk Minc, Permanents, *Amer. Math. Monthly* **72** (1965), 577–591.

A comprehensive up-to-date survey of the theory of permanents, a list of 13 conjectures and 2 problems, and a bibliography containing 16 books and 56 papers.

104. Marvin Marcus and Morris Newman, Generalized functions of symmetric matrices, *Proc. Amer. Math. Soc.* **16** (1965), 826–830.

If S is a doubly stochastic matrix and A is a positive semi-definite symmetric matrix such that $A-S$ is nonnegative, then

$$\operatorname{per}(A) \geqslant n!/n^n.$$

1966

105. W. Beineke and F. Harary, Binary matrices with equal determinant and permanent, *Studia Sci. Math. Hungar.* **1** (1966), 179–183.

Graph-theoretical conditions for a binary matrix to have equal determinant and permanent.

106. Richard A. Brualdi, On the permanent and maximal characteristic root of a nonnegative matrix, *Proc. Amer. Math. Soc.* **17** (1966), 1413–1416.

If A is an irreducible nonnegative $n\times n$ matrix with maximal characteristic root r, and if $\operatorname{per}(A)>0$, then

$$\lim_{m\to\infty} (\operatorname{per}(A^m))^{1/m} = r^n.$$

107. Richard A. Brualdi, Permanent of the direct product of matrices, *Pacific J. Math.* **16** (1966), 471–482.

Let A and B be nonnegative $m\times m$ and $n\times n$ matrices, respectively. Then

$$(\operatorname{per}(A))^n(\operatorname{per}(B))^m \leqslant \operatorname{per}(A\otimes B) \leqslant K_{m,n}(\operatorname{per}(A))^n(\operatorname{per}(B))^m,$$

where $K_{m,n}$ is a number depending on m and n only.

108. Richard A. Brualdi, Permanent of the product of doubly stochastic matrices, *Proc. Cambridge Philos. Soc.* **62** (1966), 643–647.

When is $\operatorname{per}(AB)=\operatorname{per}(A)\operatorname{per}(B)$? Gives the answer for doubly stochastic matrices, and for nonnegative matrices one of which is fully indecomposable and has a positive permanent.

109. R. A. Brualdi and M. Newman, Inequalities for the permanental minors of nonnegative matrices, *Canad. J. Math.* **18** (1966), 608–615.

The sum of all order r subpermanents of an $n\times n$ nonnegative row substochastic matrix cannot exceed $\binom{n}{r}$.

ISBN 0-201-13505-1

110. R. A. Brualdi and M. Newman, Proof of a permanental inequality, *Quart. J. Math. Oxford Ser.* (2) **17** (1966), 234–238.

If the spectral radius of a nonnegative matrix A does not exceed 1, then $\text{per}(I-A) \geqslant 0$.

111. Peter M. Gibson, A short proof of an inequality for the permanent function, *Proc. Amer. Math. Soc.* **17** (1966), 535–536.

A short proof of the inequality in [110] for substochastic matrices, using Ryser's expansion in [87].

112. W. B. Jurkat and H. J. Ryser, Matrix factorizations of determinants and permanents, *J. Algebra* **3** (1966), 1–27.

A remarkable formula expressing the permanent of a matrix as a product of matrices, each formed from only one row of the matrix. The formula is exploited to obtain bounds for permanents.

113. A. Lempel and I. Cederbaum, Minimum feedback arc and vertex sets of a directed graph, *IEEE Trans. Circuit Theory* **CT-13** (1966), 399–403.

The determination of minimum sets involves the expansion of a permanent.

114. Elliott H. Lieb, Proofs of some conjectures on permanents, *J. Math. Mech.* **16** (1966), 127–134.

If A is a positive semi-definite hermitian and $A=\begin{bmatrix} A_{11} & A_{12} \\ A_{21} & A_{22} \end{bmatrix}$, where A_{11}, A_{22} are square, then

$$\text{per}(A) \geqslant \text{per}(A_{11})\,\text{per}(A_{22}).$$

(Conjecture 8 [103]). Also proves Conjecture 9 [103] when $m=2$.

115. Marvin Marcus, Permanents of direct products, *Proc. Amer. Math. Soc.* **17** (1966), 226–231.

If A and B are complex $n \times n$ and $m \times m$ matrices, respectively, then

$$|\text{per}(A \otimes B)|^2 \leqslant (\text{per}(AA^*))^m (\text{per}(B^*B))^n.$$

If A and B are positive semi-definite hermitian, then

$$\text{per}(A \otimes B) \geqslant \left(\frac{1}{n!}\right)^m \left(\frac{1}{m!}\right)^n (\text{per}(A))^m (\text{per}(B))^n.$$

116. Marvin Marcus and Henryk Minc, A permanental inequality—the case of equality, *Canad. J. Math.* **18** (1966), 1085–1090.

A necessary and sufficient condition that $|\text{per}(A)| = n^{-1}\Sigma_{j=1}^n |\alpha_j|^n$, where A is a normal matrix with eigenvalues $\alpha_1, \ldots, \alpha_n$.

117. Marvin Marcus and Henryk Minc, *Modern University Algebra*, New York, 1966.

Contains some material on permanents: elementary properties and the Dance Problem.

118. L. Mirsky and Hazel Perfect, Systems of representatives, *J. Math. Anal. Appl.* **15** (1966), 520–568.

ISBN 0-201-13505-1

An extensive survey of results on systems of representatives. Many results involving finite sets can be stated in terms of permanents, which are not explicitly mentioned in the paper.

119. Herbert S. Wilf, On the permanent of a doubly stochastic matrix, *Canad. J. Math.* **18** (1966), 758–761.

Estimates and asymptotic results for the average of $\text{per}(\Sigma_{i=1}^{s}P_i/s)$ as the P_i run independently over permutation matrices.

1967

120. Leonard E. Baum and J. A. Eagon, An inequality with applications to statistical estimation for probabilistic functions of Markov processes and to a model for ecology, *Bull. Amer. Math. Soc.* **73** (1967), 360–363.

The inequality when applied to permanents yields the following result. Let S be the set of $m\times n$ row stochastic matrices with nonzero permanents. Let $f: S\to S$ be the function defined by

$$(f(A))_{ij}=a_{ij}\,\text{Per}(A(i|j))/\text{Per}(A),$$

$i=1,\ldots,m, j=1,\ldots,n$, for all $A=(a_{ij})\in S$. Then $\text{Per}(f(A))>\text{Per}(A)$, unless $f(A)=A$.

121. J. L. Brenner and R. A. Brualdi, Eigenschaften der Permanentefunktion, *Arch. Math.* **18** (1967), 585–586.

If A is a substochastic matrix, then all the roots of $\text{per}(zI-A)$ lie within the disc $0\leqslant|z|\leqslant 1$. (Conjecture 13 in [103].)

122. Peter Botta, Linear transformations that preserve the permanent, *Proc. Amer. Math. Soc.* **18** (1967), 566–569.

A new proof of a theorem of Marcus and May [76].

123. D. Ž. Đoković, On a conjecture by van der Waerden, *Mat. Vesnik* (19) **4** (1967), 272–276.

Let $p_k(A)$ denote the sum of order k subpermanents of A. Conjecture: If $A\in\Omega_n$, $1\leqslant k\leqslant n$, then

$$p_k(A)\geqslant\frac{(n-k+1)^2}{nk}\,p_{k-1}(A).$$

The conjecture is proved for $k=3$.

124. D. Ž. Đoković, Some permanental inequalities, *Publ. Inst. Math.* (*Beograd*) (N.S.) (21) **7** (1967), 191–195.

If A and B are $m\times n$ and $n\times k$ matrices, and $\alpha\in G_{r,m}$, $\beta\in G_{r,k}$, then

$$\text{per}((AB)[\alpha|\beta])^2\leqslant\text{per}((AA^*)[\alpha|\alpha])\,\text{per}((B^*B)[\beta|\beta]).$$

If A is a positive semi-definite doubly stochastic $n\times n$ matrix and $\alpha\in G_{r,n}$, then $\text{per}(A[\alpha|\alpha])\geqslant r!/n^r$.

ISBN 0-201-13505-1

125. W. H. Greub, *Multilinear Algebra*, Berlin, 1967.
Contains an expression for the permanent as an inner product of symmetrized tensors.

126. W. B. Jurkat and H. J. Ryser, Term ranks and permanents of nonnegative matrices, *J. Algebra* **5** (1967), 342–357.
Let A be a nonnegative matrix and let $r_1' \leqslant \cdots \leqslant r_n'$, $c_1' \leqslant \cdots \leqslant c_n'$ denote the row sums and the column sums arranged in ascending order. Then

$$\operatorname{per}(A) \leqslant \prod_{j=1}^{n} \min(r_j', c_j').$$

127. Marvin Marcus, Lengths of tensors, *Inequalities* (*Proc. Sympos. Wright-Patterson Air Force Base, Ohio* 1965) 163–176, New York, 1967.
If $A = (a_{ij})$ is an $n \times n$ positive definite hermitian matrix whose minimum eigenvalue is λ_n, then

$$\operatorname{per}(A) \geqslant \prod_{i=1}^{n} a_{ii} + \lambda_n^{n-2}\Bigl(\sum_{n>j>i>1} |a_{ij}^2|\Bigr).$$

128. Marvin Marcus and Henryk Minc, On a conjecture of B. L. van der Waerden, *Proc. Cambridge Philos. Soc.* **63** (1967), 305–309.
Conjecture: If S is a doubly stochastic $n \times n$ matrix, $n \geqslant 2$, then

$$\operatorname{per}(S) \geqslant \operatorname{per}\Bigl(\frac{J-S}{n-1}\Bigr).$$

It is shown that the conjecture is true in a sufficiently small neighborhood of J_n, and also for positive semi-definite S. The conjecture implies the van der Waerden conjecture.

129. Marvin Marcus and Henryk Minc, Permutations on symmetry classes, *J. Algebra* **5** (1967), 59–71.
A is an $n \times n$ generalized permutation matrix if and only if to each $\gamma \in G_{m,n}$, for some $m < n$, corresponds exactly one $\gamma \in G_{m,n}$ such that $\operatorname{per}(A[\omega|\gamma]) \neq 0$.

130. Marvin Marcus and George W. Soules, Some inequalities for combinatorial matrix functions, *J. Combinatorial Theory* **2** (1967), 145–163.
Improves inequalities of Marcus [82] and of Lieb [114].

131. Henryk Minc, A lower bound for permanents of (0,1)-matrices, *Proc. Amer. Math. Soc.* **18** (1967), 1128–1132.
A simple proof of an inequality obtained in [112]: If A is a (0,1)-matrix with row sums $r_1 \geqslant \cdots \geqslant r_n$, then

$$\operatorname{per}(A) \geqslant \prod_{i=1}^{n} \max(r_i + i - n, 0).$$

The case of equality is discussed.

132. Henryk Minc, An inequality for permanents of (0,1)-matrices, *J. Combinatorial Theory* **2** (1967), 321–326.

ISBN 0-201-13505-1

Improves a result in [84]: If A is a (0,1)-matrix with row sums $r_1,\dots,r_n$, then

$$\operatorname{per}(A) \leqslant \prod_{i=1}^{n} \frac{r_i+\sqrt{2}}{1+\sqrt{2}}.$$

133. T. S. Motzkin, Signs of minors, *Inequalities* (*Proc. Sympos. Wright-Patterson Air Force Base, Ohio* 1965) 225–240, New York, 1967.

Generalizes results on matrices with dominant main diagonal, including a result on permanents.

134. D. W. Sasser and M. L. Slater, On the inequality $\Sigma x_i y_i \geqslant (1/n)\Sigma x_i \cdot \Sigma y_i$ and the van der Waerden conjecture, *J. Combinatorial Theory* **3** (1967), 25–33.

If A is a normal doubly stochastic $n\times n$ matrix, $A \neq J_n$, and if all eigenvalues of A lie in the wedge defined by $|\arg z| \leqslant \pi/2k$, for some k, $2 \leqslant k \leqslant n$, then the sum of all subpermanents of A of order k is greater than that of J_n.

1968

135. L. E. Baum and G. R. Sell, Growth transformations for functions on manifolds, *Pacific J. Math.* **27** (1968), 211–227.

Improves the result in [120]:

$$\operatorname{Per}(A) \leqslant \operatorname{Per}((1-t)A + tf(A)),$$

for any $0 < t \leqslant 1$. Equality holds if and only if $f(A) = A$.

136. L. B. Beasley and J. L. Brenner, Bounds for permanents, determinants, and Schur functions, *J. Algebra* **10** (1968), 134–148.

Upper and lower bounds for the permanent of a matrix with dominant main diagonal.

137. Peter Botta, On the conversion of the determinant into the permanent, *Canad. Math. Bull.* **11** (1968), 31–34.

Re-proves the result in [70].

138. Henry H. Crapo, Permanents by Möbius inversion, *J. Combinatorial Theory* **4** (1968), 198–200.

Möbius inversion techniques are used to establish Ryser's formula for evaluation of permanents, and to establish an alternative method.

139. P. J. Eberlein and Govind S. Mudholkar, Some remarks on the van der Waerden conjecture, *J. Combinatorial Theory* **5** (1968), 386–396.

Two elementary proofs of the van der Waerden conjecture for the case $n=3$, and a proof for the case $n=4$.

140. P. Erdős and A. Rényi, On random matrices. II, *Studia Sci. Math. Hungar.* **3** (1968), 459–464.

Conjecture: $\liminf(\operatorname{per}(A))^{1/n} > 1$ for A in Λ_n^k and fixed $k \geqslant 3$.

ISBN 0-201-13505-1

141. P. M. Gibson, An inequality between the permanent and the determinant, *Proc. Amer. Math. Soc.* **19** (1968), 971–972.
If A is a substochastic matrix, then

$$\operatorname{per}(I-A) \geqslant \det(I-A) \geqslant 0.$$

142. J. M. Hammersley, An improved lower bound for the multidimensional dimer problem, *Proc. Cambridge Philos. Soc.* **64** (1968), 455–463.
Uses permanents to obtain a lower bound for the dimer problem. An improved bound is obtained if the van der Waerden conjecture is assumed.

143. V. Hval, On the permanent of a space matrix (Russian), *Mat. Vesnik* (20) **5** (1968), 173–176.
An identity for the permanent of a space matrix.

144. Marvin Marcus and Henryk Minc, Extensions of classical matrix inequalities, *Linear Algebra and Appl.* **1** (1968), 421–444.
Lower bounds for the permanents and for the sums of subpermanents of order m of normal doubly stochastic matrices. In particular, a result of Sasser and Slater [134] is improved.

145. Marvin Marcus and Stephen Pierce, On a combinatorial result of R. A. Brualdi and M. Newman, *Canad. J. Math.* **20** (1968), 1056–1067.
A permanental inequality of Brualdi and Newman [109] is obtained as a particular case of an inequality for certain Schur functions.

146. J. M. S. Simões Pereira, Boolean permanents, permutation graphs and products, *SIAM J. Appl. Math.* **16** (1968), 1251–1254.
Boolean permanents are used to study the existence of permutation subgraphs.

147. Herbert S. Wilf, A mechanical counting method and combinatorial applications, *J. Combinatorial Theory* **4** (1968), 246–258.
A generating function for the principle of inclusion and exclusion is applied to obtain an upper bound for the permanents of nonnegative matrices and an upper bound for the permanents of (0, 1)-matrices.

148. Herbert S. Wilf, Divisibility properties of the permanent function, *J. Combinatorial Theory* **4** (1968), 194–197.
The permanent of the incidence matrix of a (v,k,λ) configuration, where k and λ are even, and $v \geqslant 3$ is divisible by $2^{[(v+3)/4]}$.

1969

149. Ali R. Amir-Moéz, *An Introduction to Elements of Multilinear Algebra*, Lubbock, 1969.
Contains some material on permanents of square matrices as an inner product on symmetric spaces.

150. Leroy B. Beasley, Maximal groups on which the permanent is multiplicative, *Canad. J. Math.* **21** (1969), 495–497.

ISBN 0-201-13505-1

Answers in the affirmative a question of Marcus and Minc [103]: The set of all $n\times n$ generalized permutation matrices is a maximal group of $n\times n$ nonsingular matrices on which the permanent is multiplicative.

151. D. Ž. Đoković, Simple proof of a theorem on permanents, *Glasgow Math. J.* **10** (1969), 52–54.

A proof of a theorem of Lieb [114].

152. P. J. Eberlein, Remarks on the van der Waerden conjecture II, *Linear Algebra and Appl.* **2** (1969), 311–320.

A proof of the van der Waerden conjecture for the case $n=5$. Also: If $A\in\Omega_n$ and $\operatorname{per}(A)\leqslant\operatorname{per}(S)$ for all $S\in\Omega_n$, then any two rows (columns) of A are either equal or have different zero patterns.

153. P. M. Gibson, An identity between permanents and determinants, *Amer. Math. Monthly* **76** (1969), 270–271.

Let $A=(a_{ij})$ and $B=(b_{ij})$ be $n\times n$ matrices. If $a_{ij}=0$ whenever $j>i+1$, and if $b_{ij}=a_{ij}$ for $i\geqslant j$, and $b_{ij}=-a_{ij}$ for $i<j$, then $\operatorname{per}(A)=\det(B)$.

154. Frank Harary, Determinants, permanents and bipartite graphs, *Math. Mag.* **42** (1969), 146–148.

Permanents are used to find the number of linear subgraphs of a directed graph and the number of 1-factors of a bipartite graph.

155. Bruce W. King and Francis D. Parker, A Fibonacci matrix and the permanent function, *Fibonacci Quart.* **7** (1969), 539–544.

Another proof of the formula for the permanent of the circulant I_n+P+P^2 [94]. Also:

$$\operatorname{per}(I_n+P+P^2)=2+\sum_{k=0}^{[n/2]}\binom{n-k}{k}+\sum_{k=0}^{[(n-2)/2]}\binom{n-k-2}{k}.$$

156. Marvin Marcus, Inequalities for matrix functions of combinatorial interest, *SIAM J. Appl. Math.* **17** (1969), 1023–1031.

A survey containing many results on permanents.

157. Marvin Marcus, Spectral properties of higher derivations on symmetry classes of tensors, *Bull. Amer. Math. Soc.* **75** (1969), 1303–1307.

Let A be an $n\times n$ normal matrix with eigenvalues $\lambda_1,\ldots,\lambda_n$, and let B be a principal $p\times p$ submatrix of A. Then the sum of all principal subpermanents of B of order r is in the convex hull of the numbers $E_r(\lambda_{\alpha_1},\ldots,\lambda_{\alpha_p})$, $\alpha\in G_{p,n}$.

158. Marvin Marcus, Subpermanents, *Amer. Math. Monthly* **76** (1969), 530–533.

Let $P_k(X)$ denote the sum of all subpermanents of X of order k. If A and B are any n-square complex matrices, then

$$|P_k(AB)|\leqslant P_k(AA^*)^{1/2}P_k(B^*B)^{1/2}.$$

159. Marvin Marcus and Paul J. Nikolai, Inequalities for some monotone functions, *Canad. J. Math.* **21** (1969), 485–494.

ISBN 0-201-13505-1

If A and B are positive semi-definite hermitian matrices, then

$$\operatorname{per}(A+B) \geqslant \operatorname{per}(A).$$

160. N. Metropolis, M. L. Stein, and P. R. Stein, Permanents of cyclic (0,1) matrices, *J. Combinatorial Theory* **7** (1969), 291–321.
Recurrence formulas for the permanents of n-square circulants $I+P+\cdots+P^{k-1}$ for $k \leqslant 6$ (for $k \leqslant 4$ see [94]). Also results for $k=7,8,9$. Numerical tables for the range $4 \leqslant k \leqslant 9$, $n \leqslant 80$.

161. Henryk Minc, Bounds for permanents of nonnegative matrices, *Proc. Edinburgh Math. Soc.* (2) **16** (1968/69), 233–237.
Bounds for permanents of nonnegative matrices of Jurkat and Ryser [112] are proved by a simple method. Improved bounds are obtained.

162. Henryk Minc, On lower bounds for permanents of (0,1) matrices, *Proc. Amer. Math. Soc.* **22** (1969), 117–123.
If $A=(a_{ij})$ is a fully indecomposable (0,1)-matrix, then

$$\operatorname{per}(A) \geqslant \left(\sum_{i,j} a_{ij}\right) - 2n + 2.$$

163. Paul J. Nikolai, Mean value and limit theorems for generalized matrix functions, *Canad. J. Math.* **21** (1969), 982–991.
Answers in the negative a question raised by Marcus and Newman [79]: The function $(\operatorname{per}(A^t))^{1/t}$, where A is positive-definite, is not, in general, convex-concave relative to any value of t.

164. P. E. O'Neil, Asymptotics and random matrices with row-sum and column-sum restrictions, *Bull. Amer. Math. Soc.* **75** (1969), 1276–1282.
Asymptotic formulas for the permanent of a sparse (0,1)-matrix, and for the average permanent in a set of sparse (0,1)-matrices with all row sums and column sums equal.

165. J. K. Percus, *Combinatorial Methods*, New York, 1969.
Generating functions for permanents. Uses permanents in various enumerative problems, including the Dimer Problem.

166. Szilveszter Rochlitz, Über die Permanente des verallgemeinerten tensoriellen Produkts der Matrizen, *Publ. Math. Debrecen* **16** (1969), 229–237.
The permanent of the generalized tensor product of two sets of nonnegative matrices cannot be smaller than the product of the permanents of the matrices.

167. H. J. Ryser, Permanents and systems of distinct representatives, *Combinatorial Mathematics and Its Applications* (*Proc. Conf. University of North Carolina*, 1967), Chapel Hill, 1969.
A survey of combinatorial properties of permanents with emphasis on unsettled questions.

168. D. W. Sasser and M. L. Slater, On a generalization of the van der Waerden conjecture, *Portugal. Math.* **28** (1969), 91–95.

ISBN 0-201-13505-1

Let $P_k(A)$ denote the sum of all subpermanents of A of order k. The only local extremum of $P_k(A)$, $2 \leqslant k \leqslant n$, in the interior of Ω_n is at J_n.

169. Richard Sinkhorn, Concerning a conjecture of Marshall Hall, *Proc. Amer. Math. Soc.* **21** (1969), 197–201.

If A is a (0, 1)-matrix with three 1's in each row and each column, then $\text{per}(A) \geqslant n$. This proves a conjecture of Marshall Hall.

170. Richard Sinkhorn and Paul Knopp, Problems involving diagonal products in nonnegative matrices, *Trans. Amer. Math. Soc.* **136** (1969), 67–75.

Introduces a canonical form for nearly decomposable nonnegative matrices.

1970

171. Leroy B. Beasley, Maximal groups on which the permanent is multiplicative: corrigendum, *Canad. J. Math.* **22** (1970), 192.

A correction to [150] as regards the uniqueness of the group.

172. Richard A. Brualdi and Morris Newman, An enumeration problem for a congruence equation, *J. Res. Nat. Bur. Standards* **74B** (1970), 37–40.

Gives the number of distinct terms in the expansion of the permanent of a circulant with (commuting) indeterminate entries.

173. A. de Luca, L. M. Ricciardi, and R. Vasudevan, Note on relating pfaffians and hafnians with determinants and permanents, *J. Mathematical Phys.* **11** (1970), 530–535.

An expression for a permanent as a sum of two hafnians.

174. G. N. de Oliveira, A conjecture and some problems on permanents, *Pacific J. Math.* **32** (1970), 495–499.

Existence of matrices with prescribed main diagonal entries and prescribed permanental eigenvalues. A conjecture about the permanental eigenvalues of an irreducible doubly stochastic matrix.

175. J. E. H. Elliott, On matrices with a restricted number of diagonal values, *Pacific J. Math.* **35** (1970), 79–82.

Proves a conjecture of Marcus [103, Conjecture 10].

176. P. M. Gibson, Combinatorial matrix functions and 1-factors of graphs, *SIAM J. Appl. Math.* **19** (1970), 330–333.

An inequality between the hafnian and the permanent is used to obtain an upper bound for the number of 1-factors of a graph.

177. Andrew M. Gleason, Remarks on the van der Waerden permanent conjecture, *J. Combinatorial Theory* **8** (1970), 54–64.

Proposes, from probabilistic arguments, a conjecture about stochastic $n \times r$ matrices that is equivalent to the van der Waerden conjecture. Proves the latter for $n=4$.

178. D. J. Hartfiel, A simplified form for nearly reducible and nearly decomposable matrices, *Proc. Amer. Math. Soc.* **24** (1970), 388–393.

Gives a simplified canonical form for nearly decomposable matrices. If A is an n-square (0, 1)-matrix having exactly three 1's in each row and column, then $\text{per}(A) \geqslant n+3$.

ISBN 0-201-13505-1

179. Mark Hedrick and Richard Sinkhorn, A special class of irreducible matrices—the nearly reducible matrices, *J. Algebra* **16** (1970), 143–150.

Nearly reducible matrices are introduced. If A is a nearly reducible (0,1)-matrix, then $\mathrm{per}(A) \leqslant 1$.

180. Marvin Marcus, Inequalities for submatrices, *Inequalities* II (*Proc. Second Sympos. U.S. Air Force Acad. Colo.*, 1967), 223–240, New York, 1970.

Let $\rho_r(X)$ denote the sum of all principal subpermanents of X of order r. If A and B are commuting n-square positive definite hermitian matrices, $\alpha \in Q_{m,n}$ and $1 \leqslant r \leqslant m \leqslant n$, then

$$\rho_r\big((A+B)^{1/r}[\alpha]\big) \geqslant \rho_r(A^{1/r}[\alpha]) + \rho_r(B^{1/r}[\alpha]).$$

Also, inequalities for $\rho_r(A)$, where A is normal.

181. Marvin Marcus and William R. Gordon, The structure of bases in tensor spaces, *Amer. J. Math.* **92** (1970), 623–640.

Uses permanents in proofs on the existence of A-normal bases.

182. Albert Nijenhuis and Herbert S. Wilf, On a conjecture in the theory of permanents, *Bull. Amer. Math. Soc.* **76** (1970), 738–739.

Announcement of the following result: If A is an n-square (0,1)-matrix with row sums $r_1,\ldots,r_n$, then

$$\mathrm{per}(A) \leqslant \prod_{i=1}^{n} \big((r_i!)^{1/r_i} + \tau\big),$$

where $\tau = 0.1367\ldots$ is a universal constant.

183. Albert Nijenhuis and Herbert S. Wilf, On a conjecture of Ryser and Minc, *Nederl. Akad. Wetensch. Proc. Ser. A* 73 = *Indag. Math.* **32** (1970), 151–157.

A proof of the result in [182].

184. Patrick Eugene O'Neil, Asymptotics in random (0,1)-matrices, *Proc. Amer. Math. Soc.* **25** (1970), 290–296.

The asymptotic behavior of the average value of the permanent, for all $n \times n$ (0,1)-matrices, as a function of n.

185. Phillip A. Ostrand, Systems of distinct representatives II, *J. Math. Anal. Appl.* **32** (1970), 1–4.

Let A be an $m \times n$ (0,1)-matrix with row sums $r_1 \leqslant \cdots \leqslant r_m$ and with a positive permanent. Then

$$\mathrm{Per}(A) \leqslant \prod_{i=1}^{n} \max(1, r_i - i + 1).$$

186. Edwin J. Roberts, *The fully indecomposable matrix and its associated bipartite graph—an investigation of combinatorial and structural properties*, Ph.D. Dissertation, NASA Technical Memorandum, TM X-58037, 1970.

A lower bound estimate for the permanents of fully indecomposable (0,1)-matrices, and partial results on upper bounds for the permanents of nearly decomposable (0,1)-matrices.

ISBN 0-201-13505-1

187. Henry Sharp Jr., The permanent of a transitive relation, *Proc. Amer. Math. Soc.* **26** (1970), 153–157.

Permanents of incidence matrices for transitive relations.

1971

188. Morton Abramson, A note on permanents, *Canad. Math. Bull.* **14** (1971), 1–4.

A generalization of permanents of rectangular matrices, with some combinatorial applications.

189. J. Csima, A class of counterexamples on permanents, *Pacific J. Math.* **37** (1971), 655–656.

Counter-examples to a conjecture of de Oliveira [174] on permanental eigenvalues of a doubly stochastic matrix.

190. G. N. de Oliveira, Note on the function $\operatorname{per}(\lambda I - A)$, *Univ. Lisboa Revista Fac. Ci. A* (2) **13** (1970/71), 199–201.

A simple proof of the result in [174].

191. P. Erdös, Problems and results in combinatorial analysis, *Combinatorics, Proc. Sympos. in Pure Math.*, **XIX**, 77–89, 1971.

Contains a statement of the van der Waerden conjecture.

192. P. M. Gibson, Conversion of the permanent into the determinant, *Proc. Amer. Math. Soc.* **27** (1971), 471–476.

Let A be an n-square $(0,1)$-matrix with positive permanent. If the permanent of A can be converted into a determinant by affixing $\pm$ signs to the entries of A, then A has at most $(n^2+3n-2)/2$ positive entries.

193. V. L. Girko, Inequalities for a random determinant and a random permanent, *Teor. Verojatnost. i Mat. Statis. Vyp.* **4** (1971), 48–57.

Inequalities connecting the probability distribution of $\operatorname{per}(A)$, where A is a matrix whose entries are independent random variables, with certain characteristic functions and with Laplace transforms.

194. Michael Lee Graf, *On the van der Waerden conjecture—an automated approach*, M.S. thesis, Wright State University, 1971.

Determines certain regions in Ω_n, for $n=6,7,8$, in which the van der Waerden conjecture holds.

195. D. J. Hartfiel, Counterexamples to a conjecture of G. N. de Oliveira, *Pacific J. Math.* **38** (1971), 67–68.

Counter-examples to a conjecture of de Oliveira [174].

196. D. J. Hartfiel and J. W. Crosby, On the permanent of a certain class of $(0,1)$-matrices, *Canad. Math. Bull.* **14** (1971), 507–511.

If A is an n-square $(0,1)$-matrix with three 1's in each row and each column, then $\operatorname{per}(A) \geqslant 3(n-1)$.

197. Roy B. Levow, A graph theoretic solution to a problem of Marshall Hall, *Proc. Second Louisiana Conf. on Combinatorics, Graph Theory and Computing* (1971), 367–374.

ISBN 0-201-13505-1

Let A be an $n\times n$ matrix, $n\geqslant 2$, with nonnegative integer entries. If all the row sums and the column sums of A equal k, then

$$\operatorname{per}(A)\geqslant\binom{k}{2}n/3+k.$$

198. David London, On matrices with doubly stochastic pattern, *J. Math. Anal. Appl.* **34** (1971), 648–652.
Let A be an $n\times n$ nonnegative matrix with doubly stochastic pattern and row sums $r_1,\ldots,r_n$. There exist diagonal matrices $D_i=\operatorname{diag}(d_1^{(i)},\ldots,d_n^{(i)})$, $i=1,2$, such that D_1AD_2 is doubly stochastic and

$$\prod_{i=1}^{n} d_i^{(1)}d_i^{(2)}\geqslant\prod_{i=1}^{n} r_i^{-1}.$$

199. David London, Some notes on the van der Waerden conjecture, *Linear Algebra and Appl.* **4** (1971), 155–160.
Let $\operatorname{per}(A)=\min_{S\in\Omega_n}(\operatorname{per}(S))$. Then: (1) $\operatorname{per}(A(i|j))\geqslant\operatorname{per}(A)$; (2) if A is not positive, then the rows (columns) of A are of at least three different zero patterns.

200. Marvin Marcus, Linear transformations on matrices, *J. Res. Nat. Bur. Standards*, **75B** (1971), 107–113.
A survey. Quotes results in [29], [70], [76], and [122].

201. Clifford W. Marshall, *Applied Graph Theory*, New York, 1971.
Contains Ryser's formula for the evaluation of permanents. Uses permanents to enumerate SDRs.

202. Henryk Minc, Rearrangements, *Trans. Amer. Math. Soc.* **159** (1971), 497–504.
Improves bounds in [112] for permanents of (0,1)-matrices.

203. Stephen Pierce, Multilinear functions of row stochastic matrices, *Canad. J. Math.* **23** (1971), 833–843.
Generalizes two inequalities of Brualdi and Newman [109] to a class of functions that includes the permanent, and determines cases of equality that apply to the Brualdi–Newman results.

1972

204. L. B. Beasley and L. Cummings, Permanent groups, *Proc. Amer. Math. Soc.* **34** (1972), 351–355.
A characterization of groups of nonsingular matrices on which the permanent function is multiplicative.

205. G. N. de Oliveira, On the multiplicative inverse eigenvalue problem, *Canad. Math. Bull.* (2) **15** (1972), 189–193.
Some results on the inverse permanental eigenvalue problem.

206. P. M. Gibson, A lower bound for the permanent of a (0,1)-matrix, *Proc. Amer. Math. Soc.* **33** (1972), 245–246.

ISBN 0-201-13505-1

Let A be a fully indecomposable n-square (0, 1)-matrix with at least k 1's in each row. Then

$$\operatorname{per}(A) \geqslant \sigma(A) - 2n + 2 + \sum_{m=1}^{k} (m! - 1),$$

where $\sigma(A)$ denotes the sum of the entries of A.

207. P. M. Gibson, Localization of the zeros of the permanent of a characteristic matrix, *Proc. Amer. Math. Soc.* **31** (1972), 18–20.
Shows that many bounds for the eigenvalues of a complex matrix A are also bounds for the zeros of $\operatorname{per}(\lambda I - A)$.

208. D. J. Hartfiel and J. W. Crosby, A lower bound for the permanent on $U_n(k,k)$, *J. Combinatorial Theory* **12** (1972), 283–288.
Let A be an n-square (0, 1)-matrix with exactly k 1's in each row and column. Then

$$\operatorname{per}(A) \geqslant (k-2)(k-1)n/2.$$

209. Roy B. Levow, Lower bounds for permanents of incidence matrices, *J. Combinatorial Theory (A)* **12** (1972), 297–303.
Lower bounds for $\operatorname{per}(A) - |\det(A)|$, where A is the incidence matrix of a (v,k,λ) configuration.

210. Marvin Marcus and William R. Gordon, On projections in the symmetric power space, *Monatsh. Math.* **76** (1972), 130–134.
Case of equality for the inequality obtained in [159]: If A and B are positive semi-definite hermitian matrices, then

$$\operatorname{per}(A+B) \geqslant \operatorname{per}(A).$$

211. Marvin Marcus and Henryk Minc, An inequality for Schur functions, *Linear Algebra and Appl.* **5** (1972), 19–28.
If A_1 is a positive definite matrix, A_2 a positive semi-definite matrix, and μ_1, μ_2 complex numbers, then

$$|\operatorname{per}(\mu_1 A_1 + \mu_2 A_2)| \leqslant \operatorname{per}(|\mu_1| A_1 + |\mu_2| A_2).$$

212. R. Merris and S. Pierce, Monotonicity of positive semidefinite hermitian matrices, *Proc. Amer. Math. Soc.* **31** (1972), 437–440.
If A, B and $A-B$ are positive semi-definite hermitian matrices and $\operatorname{per}(A) = \operatorname{per}(B) \neq 0$, then $A = B$.

213. Henryk Minc, On permanents of circulants, *Pacific J. Math.* **42** (1972), 477–484.
Formulas for the permanent of circulant $\alpha I_n + \beta P + \gamma P^2$. In the set of doubly stochastic circulants of the form $\alpha I_n + \beta P + \gamma P^2$ the minimum permanent lies in the interval $(1/2^n, 1/2^{n-1}]$.

214. Henryk Minc, Nearly decomposable matrices, *Linear Algebra and Appl.* **5** (1972), 181–187.

ISBN 0-201-13505-1

The sum of elements in a nearly decomposable n-square (0, 1)-matrix cannot exceed $3(n-1)$. If A is a fully indecomposable (0, 1)-matrix with row sums $r_1,\dots,r_n$ and column sums $c_1,\dots,c_n$, then

$$\operatorname{per}(A) \geqslant \max(r_1,\dots,r_n,c_1,\dots,c_n).$$

215. Albert Nijenhuis and Herbert S. Wilf, Induced Markov chains and the permanent function, *Nederl. Akad. Wetensch. Proc. Ser. A* 75 = *Indag. Math.* **34** (1972), 93–99.
A generalization of permanents in a probabilistic setting.

216. O. S. Rothaus, Study of the permanent conjecture and some generalizations, *Bull. Amer. Math. Soc.* **78** (1972), 749–752.
Announces the following results: (1) if $A \in \Omega_n$, then $\operatorname{per}(A) \geqslant 1/n^{n-1}$; (2) there exists an r, depending on n, such that $\operatorname{per}(A^r)$, for $A \in \Omega_n$, achieves its minimum uniquely at J_n.

217. Richard Sinkhorn, Continuous dependence on A in the D_1AD_2 theorems, *Proc. Amer. Math. Soc.* **32** (1972), 395–398.
The techniques used involve permanents.

1973

218. Leroy B. Beasley and Larry Cummings, Permanent groups II, *Proc. Amer. Math. Soc.* **40** (1973), 358–364.
Generalizes the result of Beasley [150] to $n\times n$ matrices over any infinite field of characteristic 0 or greater than n.

219. L. M. Brègman, Certain properties of nonnegative matrices and their permanents, *Dokl. Akad. Nauk SSSR* **211** (1973), 27–30. (*Soviet Math. Dokl.* **14** (1973), 945–949.)
Proves the conjecture of Minc [84] on the upper bound for the permanents of (0, 1)-matrices.

220. L. J. Cummings, Elementary-maximal permanent semigroups, *Proc. 4th S.-E. Conf. Combinatorics, Graph Theory and Computing* (1973), 233–236.
On certain semigroups of triangular matrices on which the permanent is multiplicative.

221. Graciano Neves de Oliveira, *Generalized Matrix Functions*, Oeiras, 1973.
A book consisting of lecture notes on generalized matrix functions (Schur functions). Several results on permanents are included.

222. Gernot M. Engel, Regular equimodular sets of matrices for generalized matrix functions, *Linear Algebra and Appl.* **7** (1973), 243–274.
Let $A=(a_{ij})$ be an $n\times n$ complex matrix, and let $S(A)$ denote either the set of complex matrices $B=(b_{ij})$ such that $|a_{ij}|=|b_{ij}|$ for all i, j, or the set of all $B=(b_{ij})$ such that $b_{ii}=a_{ii}$ and $|b_{ij}|\leqslant|a_{ij}|$ for all i, j. Then

$$\min_{B\in S(A)}|\det(B)| \leqslant \min_{B\in S(A)}|\operatorname{per}(B)|.$$

223. Gernot M. Engel and Hans Schneider, Cyclic and diagonal products on a matrix, *Linear Algebra and Appl.* **7** (1973), 301–335.

ISBN 0-201-13505-1

A nonnegative matrix A is fully indecomposable if and only if, for all complex B, the conditions $|B| \leqslant A$ and $|\mathrm{per}(B)| = \mathrm{per}(A)$ imply that there exist diagonal complex matrices D_1, D_2 such that $D_1 B D_2 = A$.

224. Gernot M. Engel and Hans Schneider, Inequalities for determinants and permanents, *Linear and Multilinear Algebra* **1** (1973), 187–201.

Let $M = (m_{ij})$ be a real $n \times n$ matrix with $m_{ii} \geqslant 0, i = 1, \ldots, n$, and $m_{ij} \leqslant 0$ if $i \neq j$. Then

$$\det(M) + \mathrm{per}(M) \geqslant 2 \prod_{i=1}^{n} m_{ii}.$$

If $S = (s_{ij})$ is a stochastic $n \times n$ matrix such that $s_{ii} \geqslant 1/2, i = 1, \ldots, n$, then $\mathrm{per}(S) \geqslant 1/2^{n-1}$.

225. C. J. Everett and P. R. Stein, The asymptotic number of (0,1)-matrices with zero permanent, *Discrete Math.* **6** (1973), 29–34.

The number of n-square (0,1)-matrices with zero permanent is asymptotic to $n \cdot 2^{n^2 - n + 1}$.

226. R. C. Griffiths, An expansion for the permanent of a doubly stochastic matrix, *J. Austral. Math. Soc.* **15** (1973), 504–509.

An infinite series expansion for $\mathrm{per}(xA + (1-x)J_n)$, where $A \in \Omega_n$ and $0 < x \leqslant 1$.

227. Béla Gyires, Discrete distribution and permanents, *Publ. Math. Debrecen* **20** (1973), 93–106.

Some applications of permanents in probability theory. Equalities and inequalities for permanents, starting from probabilistic models. A procedure for approximative determination of the eigenvalues of positive semi-definite matrices based on computation of permanents.

228. Béla Gyires, On the permanent-derivatives of doubly stochastic matrices, *Demonstration Math.* **6** (1973), 657–663.

If the van der Waerden conjecture holds and all subpermanents of order $n-2$ of a doubly stochastic $n \times n$ matrix A are equal, then $A = J_n$.

229. D. J. Hartfiel, A lower bound on the permanent of a (0,1)-matrix, *Proc. Amer. Math. Soc.* **39** (1973), 83–85.

A lower bound for the permanent of fully indecomposable n-square (0,1)-matrices with row sums exceeding 2. Improves lower bounds in [162] and [206].

230. James J. Johnson, The permanent function and the problem of Montfort, *Math. Mag.* **46** (1973), 80–83.

The formula for $\mathrm{per}(J - I_n)$ (see, e.g., [87]) is obtained by elementary matrix methods.

231. Roy B. Levow, Counterexamples to conjectures of Ryser and de Oliveira, *Pacific J. Math.* **44** (1973), 603–606.

Counter–examples to conjectures of Ryser [68] (that the condition of commutativity in Tinsley's theorem [69] may be dropped) and of de Oliveira [174].

232. David London, On a connection between the permanent function and polynomials, *Linear and Multilinear Algebra* **1** (1973), 231–240.

Let $\alpha_1, \ldots, \alpha_n$ be complex numbers belonging to a circular region C, and let $\beta_1, \ldots, \beta_n$ belong to the complement of C. Let $M = (m_{ij})$, where $m_{ij} = \alpha_i - \beta_j, i,j = 1, \ldots, n$. Then $\mathrm{per}(M) \neq 0$. Let $A \in \Omega_n$. Then $\rho(A - J_n) = \rho(A) - 1$, where ($\rho(X)$ denotes the rank of X.

ISBN 0-201-13505-1

233. Marvin Marcus, *Finite Dimensional Multilinear Algebra, Part* I, New York, 1973.
Contains a treatment of Schur functions. Specialization to permanents is mostly relegated to exercises; some of these are sophisticated, but solutions are hinted.

234. Marvin Marcus and Russell Merris, A relation between permanental and determinantal adjoints, *J. Austral. Math. Soc.* **15** (1973), 270–271.
Let A be a positive semi-definite hermitian $n \times n$ matrix, and let $P(A), D(A)$ denote matrices whose (i,j) entries are $\mathrm{per}(A(j|i))$ and $(-1)^{i+j}\det(A(j|i))$, respectively. Then $n(\det(A))^{-1}D(A)-(\mathrm{per}(A))^{-1}P(A)$ is positive semi-definite.

235. Russell Merris, The permanent of a doubly stochastic matrix, *Amer. Math. Monthly* **80** (1973), 791–793.
Proposes the conjecture: If A is a doubly stochastic $n \times n$ matrix, then

$$n\,\mathrm{per}(A) \geqslant \min_i \sum_{j=1}^{n} \mathrm{per}(A(j|i)),$$

and gives a counter-example to the analogous inequality with *max* replacing *min*.

236. Henryk Minc, (0,1)-matrices with minimal permanents, *Israel J. Math.* **15** (1973), 27–30.
The permanent of a fully indecomposable (0,1)-matrix is equal to its largest row (column) sum if and only if all its other row (column) sums equal 2.

237. Hazel Perfect, Positive diagonals of ± 1-matrices, *Monatsh. Math.* **77** (1973), 225–240.
Some results on the existence of an n-square ± 1-matrix with exactly k positive diagonal products—that is, with permanent equal to $2k-n!$.

238. E. Seneta, *Nonnegative Matrices*, New York, 1973.
Mentions the van der Waerden permanent conjecture.

239. Richard Sinkhorn, Doubly stochastic matrices whose squares leave the permanent invariant, *Linear and Multilinear Algebra* **1** (1973), 103–118.
The equations $A^2 = PAQ^{\mathrm{T}}$ and $A^2 = PA^{\mathrm{T}}Q^{\mathrm{T}}$ are solved when A is doubly stochastic, and P and Q are permutation matrices. This gives a partial solution to the equation $\mathrm{per}(A^2) = \mathrm{per}(A)$ for A doubly stochastic.

1974

240. E. R. Caianiello, *Combinatorics and Renormalisation in Quantum Field Theory*, New York, 1974.
A survey of the work of Caianiello and his collaborators. Permanents and hafnians are applied to boson fields.

241. Jacques Dubois, A note of the van der Waerden permanent conjecture, *Canad. J. Math.* **26** (1974), 352–354.
On the impossibility of certain zero patterns for a doubly stochastic matrix with minimal permanent.

242. Thomas H. Foregger, On facial minimizing matrices for the permanent function, *Notices AMS* **21** (1974), A-434.
Let F be a subpolyhedron of Ω_n spanned by a set of permutation matrices, and suppose

ISBN 0-201-13505-1

that F contains a fully indecomposable matrix. If $A \in F$ and $\operatorname{per}(A) = \min_{X \in F} \operatorname{per}(X)$, then A is fully indecomposable.

243. Shmuel Friedland, Matrices satisfying the van der Waerden conjecture, *Linear Algebra and Appl.* **8** (1974), 521–528.

Any $n \times n$ doubly stochastic matrix whose numerical range lies in the sector from $-\pi/2n$ to $\pi/2n$ satisfies the van der Waerden conjecture. This improves the results of Marcus and Newman [79], and of Sasser and Slater [134].

244. Shmuel Gal and Yuri Breitbart, A method for obtaining all the solutions of a perfect matching problem, *IBM Israel Scientific Center Technical Report* **016** (1974).

An efficient method for computing the permanent of a (0, 1)-matrix in case the permanent is not large.

245. B. Gordon, T. S. Motzkin, and L. Welch, Permanents of 0, 1-matrices, *J. Combinatorial Theory Ser. A* **17** (1974), 145–155.

For any positive integer k there exist (0, 1)-matrices with permanent k. The minimum order of (0, 1)-matrices with permanent k does not exceed $[\log_2(k-1)]+2$ for $k = 2, 3, \ldots$.

246. R. C. Griffiths, Permanents of random doubly stochastic matrices, *Canad. J. Math.* **26** (1974), 600–607.

The van der Waerden conjecture and a generalization thereof from a probabilistic point of view.

247. D. J. Hartfiel, A lower bound for the permanent on a special class of matrices, *Canad. Math. Bull.* **17** (1974), 529–530.

If A is an n-square (0, 1)-matrix with r 1's in each row and column, then $\operatorname{per}(A) \geqslant (r-1)! + n(r-2)! + \cdots + n(2!) + n + 1$.

248. Mark Blondeau Hedrick, The permanent at a minimum on certain classes of doubly stochastic matrices, *Bull. Amer. Math. Soc.* **80** (1974), 836–838.

Slightly generalizes and re-proves a result in [62] and [199] (see also [269]).

249. James J. Jυhnson, Bounds for certain permanents and determinants, *Linear Algebra and Appl.* **8** (1974), 57–64.

A lower bound for the permanents of real matrices with nonnegative dominant main diagonal.

250. Henryk Minc, An unresolved conjecture on permanents of (0, 1)-matrices, *Linear and Multilinear Algebra* **2** (1974), 115–121.

If A is an n-square (0, 1)-matrix with row sums $r_1, \ldots, r_n, r_i \leqslant 8, i = 1, \ldots, n$, then

$$\operatorname{per}(A) \leqslant \prod_{i=1}^{n} (r_i!)^{1/r_i}.$$

251. Henryk Minc, A remark on a theorem of M. Hall, *Canad. Math. Bull.* **17** (1974), 547–548.

If A is an $m \times n$ (0, 1)-matrix, $m \leqslant n$, and if each row sum of A is greater than or equal to m, then A has a positive permanent.

252. J. W. Neuberger, Norm of symmetric product compared with norm of tensor product, *Linear and Multilinear Algebra* **2** (1974), 115–121.

ISBN 0-201-13505-1

An inequality between the tensor and the symmetric products of multilinear functions. The inequality is related to results of Lieb [114], and of Marcus and Soules [130].

253. O. S. Rothaus, Study of the permanent conjecture and some of its generalizations, *Israel J. Math.* **18** (1974), 75–96.
Defines, for any convex polytope, a function generalizing the permanent. Proves the results announced in [216].

254. Jacobus H. van Lint, *Combinatorial Theory Seminar Eidhoven University of Technology*, Berlin, 1974.
Contains material on general properties of permanents and in particular on the permanents of (0, 1)-matrices. Also, material on inequalities involving permanents, including an extensive discussion of the status of Minc's conjecture [84] (prior to Brégman's proof [219]).

255. Edward Tzu–Hsia Wang, A new class of finite cyclic permanent groups, *J. Combinatorial Theory Ser. A* **17** (1974), 261–264.
Cyclic groups of $n \times n$ matrices on which the permanent function is multiplicative.

256. Edward Tzu–Hsia Wang, Diagonal sums of doubly stochastic matrices, *Linear Algebra and Appl.* **8** (1974), 483–505.
Inequalities for the permanent of a doubly stochastic matrix involving the maximum and the minimum diagonal sums of the matrix.

257. Edward Tzu–Hsia Wang, On permanents of (1, – 1)-matrices, *Israel J. Math.* **18** (1974), 353–361.
Upper bound for $|\mathrm{per}(A)|$, where A is a $(1, -1)$-matrix. If H is an Hadamard matrix, then $|\mathrm{per}(H)| \leqslant |\det(H)|$.

258. William Watkins, Convex matrix functions, *Proc. Amer. Math. Soc.* **44** (1974), 31–34.
If A, B, and $A-B$ are positive semi-definite hermitian matrices and $0 \leqslant \lambda \leqslant 1$, then

$$\mathrm{per}(\lambda A + (1-\lambda)B) \leqslant \lambda\,\mathrm{per}(A) + (1-\lambda)\,\mathrm{per}(B).$$

259. Richard M. Wilson, Nonisomorphic Steiner triple systems, *Math. Z.* **135** (1974), 303–313.
Lower bounds for the number of Latin squares and for the number of nonisomorphic Steiner triple systems can be substantially improved if the van der Waerden conjecture is assumed to be true.

1975

260. Richard A. Brualdi and Thomas H. Foregger, Matrices with constant permanental minors, *Linear and Multilinear Algebra* **3** (1975), 227–243.
Properties of (0, 1)-matrices with constant permanental minors. Conjecture: If A is an n-square (0, 1)-matrix all of whose permanental minors of order $n-1$ have a common nonzero value, then $A = J$ or $I_n + P$.

261. D. Ž. Đjoković, On Neuberger's identity for symmetric tensors, *Linear and Multilinear Algebra* **2** (1975), 305–309.
Re-proves an identity of Neuberger [252] for symmetric tensors, and deduces an inequality of Lieb [114].

ISBN 0-201-13505-1

262. Thomas H. Foregger, An upper bound for the permanent of a fully indecomposable matrix, *Proc. Amer. Math. Soc.* **49** (1975), 319–324.

If A is an $n \times n$ fully indecomposable matrix with nonnegative integer entries, then

$$\operatorname{per}(A) \leqslant 2^{\sigma(A)-2n}+1,$$

where $\sigma(A)$ denotes the sum of the entries of A.

263. B. Gordon, T. S. Motzkin, and L. Welch, Errata: "Permanents of 0,1-matrices" (J. Combinatorial Theory Ser A 17 (1974), 145–155), *J. Combinatorial Theory Ser. A* **19** (1975), 367.

Announces that some of the results in [245] are not new.

264. D. J. Hartfiel, A canonical form for fully indecomposable (0, 1)-matrices, *Canad. Math. Bull.* **18** (1975), 223–227.

Obtains a canonical form for fully indecomposable (0, 1)-matrices and uses it to prove the result in [196].

265. Mark Blondeau Hedrick, p-minors of a doubly stochastic matrix at which the permanent achieves a minimum, *Linear Algebra and Appl.* **10** (1975), 177–179.

A combinatorial proof of a result of London [199].

266. J. R. Henderson, Permanents of (0, 1)-matrices having at most two zeros per line, *Canad. Math. Bull.* **18** (1975), 353–358.

The minimum permanent in the set of $n \times n$ matrices of zeros and ones with at most two zeros in every line is U_n if n is even, and $U_n - 1$ if n is odd, where U_n denotes the nth ménage number.

267. W. J. Leahey, W. C. Herndon, and V. T. Phan, A note on permanents, *Linear and Multilinear Algebra* **3** (1975), 193–196.

The permanent of a symmetric matrix satisfying certain conditions is equal to the square of the hafnian of the matrix.

268. Russell Merris, Two problems involving Schur functions, *Linear Algebra and Appl.* **10** (1975), 155–162.

Upper bounds for permanental eigenvalues of normal matrices. If A is hermitian, then its real permanental eigenvalues lie between the largest and the smallest eigenvalues of A.

269. Henryk Minc, Doubly stochastic matrices with minimal permanents, *Pacific J. Math.* **58** (1975), 155–157.

A simple proof of a result of London [199].

270. Henryk Minc, Subpermanents of doubly stochastic matrices, *Linear and Multilinear Algebra* **3** (1975), 91–94.

If all the subpermanents of order k of an $n \times n$ doubly stochastic matrix are equal, for some $k \leqslant n-2$, then all the entries of the matrix must be equal to $1/n$.

271. Victor A. Nicholson, Matrices with permanent equal to one, *Linear Algebra and Appl.* **12** (1975), 185–188.

A nonnegative matrix M is nilpotent if and only if $\operatorname{per}(M+I)=1$.

ISBN 0-201-13505-1

272. Albert Nijenhuis and Herbert S. Wilf, *Combinatorial Algorithms*, New York, 1975.

Contains a chapter on the permanent function which includes a survey, a variant of Ryser's formula [87], and a FORTRAN program for calculating the permanent of a square matrix.

273. W. C. Pye and Melvyn W. Jeter, A minimal permanent-like function, *Linear Algebra and Appl.* **12** (1975), 171–178.

A study of a matrix function that has properties similar to those of the permanent function.

1976

274. Thøger Bang, Matrixfunktioner som med et numerisk lille deficit viser v. d. Waerdens permanenthypothese, *Proc. Scandinavian Congress*, Turkku, 1976.

Outline of a proof that $\operatorname{per}(S) \geqslant 1/e^{n-1}$ for any $n \times n$ doubly stochastic matrix S.

275. R. L. Graham and D. H. Lehmer, On the permanent of Schur's matrix, *J. Austral. Math. Soc. Ser. A* **21** (1976), 487–497.

Investigation of the permanent of Schur's matrix $M_n = (\varepsilon^{st})$, $0 \leqslant s$, $t < n$, where $\varepsilon = \exp(2\pi i/n)$.

276. Béla Gyires, On permanent inequalities, *Coll. Math. Soc. J. Bolyai*, **18** (1976), *Combinatorics*.

If $A \in \Omega_n$, then

$$\left(\operatorname{per}(A^2) + \sqrt{\operatorname{per}(AA^*)\operatorname{per}(A^*A)}\right)/2 \geqslant n!/n^n.$$

277. Paul J. Knopp and Richard Sinkhorn, Permanents of special classes of nonnegative matrices, *Linear and Multilinear Algebra* **4** (1976), 129–136.

A study of matrices on the boundary of Ω_n that are closest to J_n in Euclidean norm.

278. David London, On derivations arising in differential equations, *Linear and Multilinear Algebra* **4** (1976), 179–189.

A solution $Y(z)$ of a matrix differential equation $Y'(z) = AY(z)$ satisfies a similar equation involving the rth induced matrix of $Y(z)$ and the first derivation associated with A operating on the rth completely symmetric space.

279. Henryk Minc, The invariance of elementary symmetric functions, *Linear and Multilinear Algebra* **4** (1976), 209–215.

Determines the form of linear transformations on matrices that preserve the permanent of each matrix and either its trace or the second elementary function of its eigenvalues.

280. Albert Nijenhuis, On permanents and the zeros of rook polynomials, *J. Combinatorial Theory Ser. A* **21** (1976), 240–244.

A study of the zeros of the rook polynomial of an $m \times n$ matrix A, $\sum_{k=0}^{\infty} r_k(A)(-x)^k$, where $r_k(A)$ denotes the sum of all permanental minors of order k of A, $r_0(A) = 1$.

281. Richard Sinkhorn, Doubly stochastic matrices whose squares decrease the permanent, *Linear and Multilinear Algebra* **4** (1976), 123–128.

If A is a doubly stochastic matrix such that $PAQ = J_{n_1} \dot{+} \cdots \dot{+} J_{n_s}$, for some permutation matrices P and Q, then $\operatorname{per}(A^2) \leqslant \operatorname{per}(A)$.

ISBN 0-201-13505-1

1977

282. Eva Achilles, Permanents of doubly stochastic matrices with fixed zero pattern, *Linear and Multilinear Algebra* **5** (1977), 63–70.

The permanent achieves a local minimum in the set described in the title at matrices closest to J_n provided that the zeros are confined to a single line.

283. Richard A. Brualdi and Peter M. Gibson, The convex polyhedron of doubly stochastic matrices: I. Applications of the permanent function, *J. Combinatorial Theory Ser. A* **22** (1977), 194–230.

The permanent function is used to determine geometric properties of Ω_n.

284. O. S. Rothaus, Study of the permanent conjecture and some of its generalizations. II, *Trans. Amer. Math. Soc.* **232** (1977), 143–154.

A sequel to [253]. A study of the properties of a function generalizing the permanent function.

285. Larry J. Cummings and Jennifer Seberry Wallis, An algorithm for the permanent of circulant matrices, *Canad. Math. Bull.* **20** (1977), 67–70.

An algorithm for computing the permanent of a circulant matrix with entries from any field.

286. Richard Sinkhorn, Doubly stochastic matrices with dominant p-minors, *Linear and Multilinear Algebra* **5** (1977), 107–117.

Conjecture. If $A \in \Omega_n$ and $\operatorname{per}(A(i|j)) \geqslant \operatorname{per}(A)$ for all i, j, then either $A = J_n$ or $\frac{1}{2}(I_n + P)$ up to permutations of lines. The conjecture is shown to be true under certain restrictions on A.

287. Edward T. H. Wang, On a conjecture of Marcus and Minc, *Linear and Multilinear Algebra* **5** (1977), 145–148.

The conjecture of Marcus and Minc [128] holds for $n = 3$.

Addenda

288. David K. Baxter, The existence of matrices with prescribed characteristic and permanental polynomials, *Linear Algebra and Appl.* (to appear).

Gives necessary and sufficient conditions for the existence of a matrix with prescribed characteristic polynomial and permanental characteristic polynomial.

289. Leroy B. Beasley and Larry J. Cummings, Permanent semigroups, *Linear and Multilinear Algebra* **5** (1978), 297–302.

Any permanent semigroup of matrices over an integral domain or over a finite field (with some restrictions on the characteristic) consists of matrices with at most one nonzero diagonal.

290. Richard A. Brualdi, On 1-factors of cubic bipartite graphs (to appear).

Let $A = (a_{ij}) \in \Lambda_n^3$, $n \geqslant 10$. Then $a_{ij} = 1$ implies that $\operatorname{per}(A(i|j)) \geqslant n$. Hence $\operatorname{per}(A) \geqslant 3n$.

ISBN 0-201-13505-1

291. Kim Ki-Hang Butler, On the van der Waerden conjecture, *J. Mat. Sci.* (to appear).
A matrix A is regular if there exists a doubly stochastic matrix X such that $A=AXA$. It is shown that regular doubly stochastic matrices satisfy the van der Waerden conjecture. (Sinkhorn [unpublished] showed that A is regular if and only if $PAQ=\Sigma_t^{\cdot} J_{n_t}$, where P and Q are permutation matrices.)

292. Thomas H. Foregger, Identities related to permanents of doubly stochastic matrices and series-reduced trees (to appear).
The van der Waerden conjecture has been proved for $n \leqslant 5$ by means of certain linear identities. It is shown that for $n \leqslant 9$ the known identities are sufficient to generate all possible linear identities.

293. Thomas H. Foregger, A note on matrices with constant permanental minors (to appear).
If all permanental minors of order t of an $m \times n$ matrix A have a common nonzero value, where $t \leqslant \min(m-2, n-1)$, then all the entries of A are equal.

294. Thomas H. Foregger, On the minimum value of the permanent of a nearly decomposable doubly stochastic matrix (to appear).
For $2 \leqslant n \leqslant 8$ the minimum value of the permanent of a nearly decomposable matrix in Ω_n is $1/2^{n-1}$.

295. Thomas H. Foregger, Remarks on a conjecture of M. Marcus and H. Minc, (to appear).
Proves a conjecture of Marcus and Minc [128] for $n=4$.

296. Shmuel Friedland, A study of the van der Waerden conjecture and its generalizations, *Linear and Multilinear Algebra* (to appear).
A study of a generalization of the van der Waerden problem. It is shown that the permanent of an $n \times n$ doubly stochastic matrix is greater than or equal to $1/n!$.

297. Shmuel Friedland and Henryk Minc, Monotonicity of permanents of doubly stochastic matrices, *Linear and Multilinear Algebra* (to appear).
For any $n \times n$ permutation matrix P the function $\operatorname{per}((1-\theta)J_n+\theta P)$ is strictly decreasing in the interval $-1/(n-1) \leqslant \theta \leqslant 0$ and strictly increasing for $0 \leqslant \theta \leqslant 1$.

298. Peter M. Gibson, Real permanental roots of doubly stochastic matrices, *Linear Algebra and Appl.* (to appear).
For each $n \geqslant 7$ there exists an $n \times n$ irreducible doubly stochastic matrix with n real permanental eigenvalues.

299. Béla Gyires, On inequalities concerning matrices of permanents, *Linear and Multilinear Algebra* **5** (1978), 279–282.
Generalizes inequalities of Marcus and Newman [79] to $m \times m$ matrices whose (i, j) entry is $\operatorname{per}(A_i A_j^{\mathrm{T}})$, where $(A_1, \ldots, A_m)$ are $n \times n$ real matrices with row and column sums equal to 1.

300. Ki Hang Kim and Fred W. Roush, On a conjecture of Erdös and Rényi, *Linear Algebra and Appl.* (to appear).
Partial results on a conjecture of Erdös and Rényi [140].

301. Henryk Minc, Evaluation of permanents, *Proc. Edinburgh Math. Soc.* (to appear).

ISBN 0-201-13505-1

Binet's formulas [1] for the permanents of $2\times n$, $3\times n$, and $4\times n$ matrices are extended to permanents of $m\times n$ matrices for any $m, m\leqslant n$.

302. Henryk Minc, An upper bound for the multidimensional dimer problem, *Math. Proc. Cambridge Philos. Soc.* (to appear).

Combines a result in [219] with the bound in [142] to obtain an improved upper bound for the multidimensional dimer problem.

303. A. Schrijver, A short proof of Minc's conjecture, *J. Combinatorial Theory, Ser. A* (to appear).

A short and elegant proof of Minc's conjecture [84], previously proved by Brégman [219].

ISBN 0-201-13505-1

Index to Bibliography

The numbers refer to items in the bibliography.

Index of Notation

The numbers refer to pages on which the symbols are defined.

NOTE. Standard notation of the theory of matrices is used in the text without explicit definitions:

$\det(A)$-determinant of A,
$\rho(A)$-rank of A,
A^T-transpose of A,
A^*-conjugate transpose of A,
T^*-adjoint of T,
I_n-$n\times n$ identity matrix,
$A_{(i)}$-ith row of A,
$A^{(j)}$-jth column of A,
$C_r(A)$-rth compound matrix of A,
$A\otimes B$-direct (Kronecker) product of A and B,
$\dim V$-dimension of V,
(u,v)-inner product of u and v,
$\langle v_1,\ldots,v_m\rangle$-subspace spanned by $v_1,\ldots,v_m$,
$\langle v_1,\ldots,v_m\rangle\perp$-orthogonal complement of $\langle v_1,\ldots,v_m\rangle$,
$v_1\wedge\cdots\wedge v_m$-Grassmann (skew-symmetric) product of $v_1,\ldots,v_m$;
etc.

Index

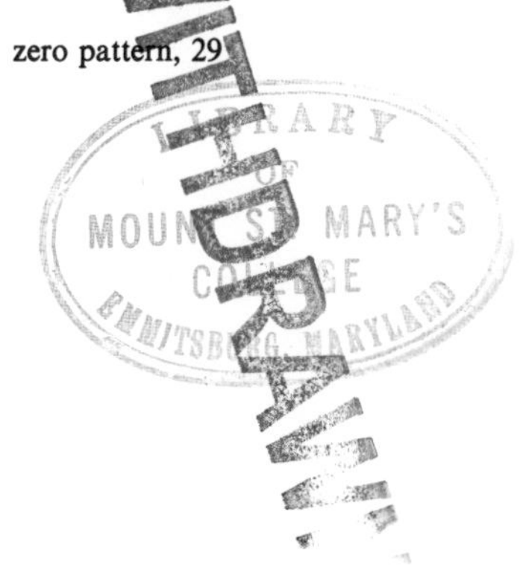